SpringerBriefs in Applied Sciences and Technology

Structural Mechanics

Series editor

Emmanuel E. Gdoutos, Department of Theoretical and Applied Mechanics, Democritus University of Thrace, Xanthi, Greece

More information about this series at http://www.springer.com/series/15039

Robert Mines

Metallic Microlattice Structures

Manufacture, Materials and Application

Springer

Robert Mines
School of Engineering
University of Liverpool
Liverpool, UK

ISSN 2191-530X   ISSN 2191-5318   (electronic)
SpringerBriefs in Applied Sciences and Technology
ISSN 2520-8020   ISSN 2520-8039   (electronic)
SpringerBriefs in Structural Mechanics
ISBN 978-3-030-15231-4   ISBN 978-3-030-15232-1   (eBook)
https://doi.org/10.1007/978-3-030-15232-1

Library of Congress Control Number: 2019933851

This Springer imprint is published by the registered company Springer Nature Switzerland AG
The registered company address is: Gewerbestrasse 11, 6330 Cham, Switzerland

Preface

The book has five main themes, namely:

1. Enhancement of specific selected structural applications using additive manufacturing lattice structures, e.g. sandwich beams, sandwich panels and energy absorption devices (static and impact), for use in lightweight structures.
2. Integrated discussion of manufacture, materials and application is followed. In the developing additive manufacturing technology, a development in one area will influence other areas, and so an integrated approach is needed.
3. The structural behaviours and applications discussed here have been studied for many years for conventional structures manufactured using traditional processes, and some of this relevant literature is discussed before introducing additive manufacture.
4. A specific set of structural cases and additively manufactured lattice structures are discussed here, but these are part of the larger fields of Architectured (Architected) Cellular Materials, and part of the larger field of Industrial Processes (including standardisation and certification).
5. Finally, an overall aim of the book is to give the researcher an introduction to some of the fundamental ideas underlying the increasingly sophisticated computer based realisation tools. These include manufacturing parameters, materials optimisation, topology optimisation and multi-function aspects.

The book is written from the perspective of the structural engineer. The book is aimed at researchers and technical practitioners, who need a focussed state of the art introduction to a specific area of additively manufactured metallic lightweight structures. In this way, the current 'design space', using state of the art technology, can be related to the fast developing technologies, and to other available additive manufacturing approaches. The approach taken here is experimental and application orientated.

The author's main research activity over the years has been in structural impact, e.g. foreign object impact and impact energy absorption. Back in 2003, the manufacturing group at the University of Liverpool was studying selective laser melting microlattice structures for bio-implants and micro-heat exchangers. These structures were identified as having potential application for use in lightweight structures. The manufacturing group at Liverpool initially worked with SLM machines from the manufacturing company MCP (2000–2006), then MTT (2006–2010) and finally Renishaw (2010–date). The selective laser melting technology has evolved from a few research laboratories in 2003 to global research activity today, with industrial products.

The discussion here, due to the author's experience, is on the first generation selective laser melting machines (MCP/MTT). First generation selective laser melting machines are discussed in the open literature, whereas, as the field becomes more industrially focused, less detail on selective laser melting machine developments is given in the open literature. Electron beam melting is discussed to compare and contrast with selective laser melting. Electron beam melting and selective laser melting are mature additive manufacturing process, and there are a number of other metal additive manufacturing processes, with more being developed all the time. Photopolymer wave guides, woven wire and binder jetting are processes that will also be discussed. Hence, the enabling additive manufacturing technology is constantly developing.

Given the fact that selective laser melting and electron beam melting are now relatively mature industrial processes, then detailed synthesis and study of realised structures can be made. The book focuses on lattice structures (including surface based lattices), as these are a well defined cellular structure that can be realised using selective laser melting, electron beam melting, photopolymer wave guides, woven wire and binder jetting. Discussion is also restricted to a few conventional materials (stainless steel, titanium alloy and aluminium alloy) and lattice topologies (Body Centred Cubic, BCCZ, Octet Truss) that are most appropriate to the selected structural applications, viz. sandwich beams and panels, and energy absorbers. Currently available additive manufacturing technology is applied to the selected structural applications, and enhancements in performances are quantified. The aim is to fully exploit additive manufacturing technologies (manufacture, materials and simulation) to enhance structural performance. The book does not push the boundaries of developing materials science for additive manufacture, but stays with conventional metallic structural materials. Currently, additive manufacture for non-metals has the greater potential for innovative development and multi functionality. However, the main potential for developing innovative metallic additive manufactured materials is at small (nano) scale.

A note on terminology. The focus of this book is on *microlattice* structures, and these are taken to have a feature size of 100–2000 µm and a cell size of 1–5 mm. Microlattices are contrasted with *nano-lattices*, the latter having feature size less than 100 nm and a cell size less than 10 µm. However, it should be noted that electroplated microlattices have wall thicknesses less than one micron. *Macro-lattices* are taken to have a feature size greater than 1 mm and a cell size greater than 5 mm, and they tend to be manufactured using conventional forming processes. These informal definitions reflect the different manufacturing processes (and to a lesser extent, materials) for each scale.

A detailed overview of the book is given in Chap. 1, and conventional lattice structural theories are given in Chap. 2. Chapter 3 discusses selected additive manufacturing processes (SLM, EBM, BJ), Chap. 4 discusses parent material and lattice characterisation, Chap. 5 discusses lattice analysis and synthesis theories, and Chap. 6 introduces photopolymer wave guides and woven wire lattice solutions. The book culminates in Chap. 7, where specific structural applications are discussed and improvements in structural performance are quantified. Finally, Chap. 8 gives conclusions and highlights future prospects, including eight suggested research themes.

The focus of the book is on overall themes and ideas, and so selected references are discussed briefly and the reader is encouraged to follow items of interest in the original papers. The book is in the form of a 'Tour d'Horizon'. The book cites over 200 journal papers, over half of which have been published within the last 4 years covering manufacturing, materials and structural applications.

Thanks are due to co-workers, Prof. Norman Jones, Wesley Cantwell and Chris Sutcliffe. Thanks are also due to graduate students and research assistants: Drs. Simon McKown, Sozon Tsopanos, Eva Shen, Recep Gümrük, Matt Smith and Rafidah Hasan. The authors' research in this field was mainly sponsored by EPSRC and EU FP6 Celpact.

Liverpool, UK   Robert Mines

Contents

Notation

A	Constant in Johnson Cook Model or Absorptivity
AlSi10/12Mg	Aluminium alloy for Additive Manufacture
AM	Additive manufacture
B	Constant in Johnson Cook model
b	Number of struts in Maxwell analysis or sandwich beam width
BCC	Body centred cubic lattice topology
BCC, Z	BCC with Z strut lattice topology
BJ	Binder jetting process
C	Constant in Johnson Cook model
CT	Computed tomography scan
D	Cowper Symonds constant or Thermal diffusivity
d	Microstrut diameter
D_i	Constraint in material for rupture
DMLS	Direct Metal Laser Sintering
E	Elastic Modulus
E^*_{BCC}	Stiffness of BCC lattice block
EBM	Electron beam melting
E_p	Laser energy in SLM process
F2 BCC	Face centred body centred cubic lattice topology
F2 CC	Face centred cubic lattice topology
FCCZ	Face centred with Z strut lattice topology
H	Specific enthalpy in SLM process
HIP	Hot iso-static processing heat treatment
j	Number of joints in Maxwell analysis
JC	Johnson Cook constitutive model
l	Gauge length or cell length
L	Length of strut
LP	Laser Power in SLM process
LR	Lloyds Register
LX	Laser exposure time in SLM process

m	State of mechanism in Maxwell analysis
M	Maxwell criterion
MIM	Metal Injection Moulding
OT	Octet truss lattice topology
PBF	Powder bed fusion process
P_f	Failure load
PPWG	Photopolymer wave guide
q	Cowper Symonds constant
s	State of self stress in Maxwell model
SLM	Selective laser melting
STL	Stereolithography file format
SMS	Size matching and scale method for optimisation
SS316L	Stainless steel 316L
t	Sandwich skin thickness
T^*	Homologous temperature in Johnson Cook model
TB	Textbook values
Ti 64	Titanium alloy Ti—6Al-4V
TPMS	Triply periodic minimal surfaces
TWI	The Welding Institute
u	Laser speed in SLM process
UV	Ultra violet light
V_{CR}	Critical impact velocity
V_i	Impact velocity
WW	Woven wire
$\dot{\varepsilon}$	Strain rate
$\dot{\varepsilon}_0$	Reference strain rate
ε_D	Densification strain
ε_f	Rupture strain
η	Stress triaxiality
ρ	Material density
$\sigma^*_{pl,BCC}$	Plastic collapse strength of BCC block
σ_0	Static block crush stress
$\sigma_{0.2}$	0.2% proof stress
σ_c	Core compression strength
σ_{CR}^{qs}	Quasi-static collapse strength
σ_d	Material dynamic yield stress
σ_s	Material static yield stress or sandwich skin tensile stress
σ_{UTS}	Ultimate tensile strength
ϕ	Laser spot diameter in SLM process

Chapter 1
Introduction and Overview

Abstract Given the wide ranging subject matter covered in this book, this chapter discusses the content and structure of the book in detail. The book takes an integrated approach to metallic microlattice structural design, as materials and manufacturing processes need to be considered in parallel with the structural realisation and application. The chapter introduces conventional structural theory, selective additive manufacturing processes, material and lattice block characterisation, theoretical methods, specialised additive manufacturing processes, structural applications (with quantified performance) and overall conclusions. The book focuses on the structural functions of sandwich construction and on energy absorption.

Keywords Metal AM processes · Metal AM materials · Sandwich construction · Energy absorption

The starting point for this book is the authors previous research work on the impact performance of sandwich beams (Mines et al. 1994), impact performance of sandwich panels (Mines et al. 1998) and the progressive collapse of cellular materials (Li et al. 2000). In the case of sandwich structures, the cellular core provides the shear stiffness (for effectiveness of the skins) and compression strength (to resist local loads). In the case of energy absorption, the behaviour sought includes progressive and stable collapse of the cellular material. Under impact loading, the parent material of the core can be strain rate dependent, and the cellular deformation and collapse modes may be inertia dependent. Traditionally, structural cellular materials include honeycomb, foams (open and closed cell), as well as natural materials (e.g. balsa wood) (Gibson and Ashby 1997). The focus here is on metallic microlattice structures.

The stiffness and strength of the sandwich core and energy absorber are dependent on the cell topology, cell scale (relative density) and the cell parent material. Loading of the cellular material may be uniaxial, or biaxial or tri-axial, with the latter including hydrostatic effects. The cellular material may be strained well into the plasticity region, and material rupture may occur. Large strain plasticity and material rupture may be dependent on structural scale.

For the cases of sandwich structures, there is an interaction between the cellular core and skins, and so deformation and failure of the sandwich structures depends on the interaction between these two. Currently the structural designer has to select from specific ranges of cellular material, with specified cell topology, cell materials and cell scale (relative density). These materials are usually constant density, with standard periodic cell. In the case of foams, the cellular topology is stochastic.

Additive manufacture provides the opportunity to realise cellular materials with extensive design freedom in cell topology, and material and cell scale (Rashed et al. 2016). A large number of additive manufacturing processes, at various scales, are being developed (Schaedler and Carter 2016). The current most relevant additive manufacturing processes for the applications considered here are selective laser melting (SLM), electron beam melting (EBM), photo polymer wave guides (PPWG), woven wire (WW) and binder jetting (BJ). These processes limit the cellular materials to a restricted range of metals, and mainly to a restricted scale (down to about 100 µm). The term microlattice is used for lattice structures with strut diameters of the order of 200 µm and cell size of the order of 2 mm. This is to differentiate with macro lattice structures (strut diameter of the order of 2+ mm and cell size of the order of 20+ mm) and nano lattices (strut diameter of the order of 200 nm and cell size of the order of 200 µm).

Hence, the focus of the book is on the realisation and on the use of cellular materials, using additive manufacturing, to enhance the impact performance of sandwich beams and panels, and energy absorbing devices. The culmination of this discussion is Chap. 7, which discusses a small number of structural solutions using current additive manufacturing techniques. To achieve these solutions, enabling technologies have to be articulated and these are discussed in Chaps. 3–6.

However, to set the scene, Chap. 2 considers appropriate structural science theories (independent of additive manufacturing) such as lattice topology, sandwich core and structural design, and impact energy absorption considerations. Lattice structures are a subset of cellular materials, and so will have advantages and disadvantages compared to other cellular materials. Core materials in sandwich structures provide a specific set of design issues, including shear stiffness and strength as well as localised stiffness and strength. Impact energy absorption requires study of micro inertia behaviour of lattice structures, and the strain rate performance of the parent material. Such impacts can be a result of blast loading, and low and high velocity foreign object impact.

Chapter 3 discusses the additive manufacturing processes of selective laser melting, electron beam melting and binder jetting, with associated material and quality issues. Selective laser melting and electron beam melting have been available since the 1990s and have recently attained a certain level of maturity. This means that the processes are becoming more mainstream, industrially, and the underlying technology is becoming more stable. However, manufacturing knowledge is becoming more closed. The approach taken here is to consider first generation selective laser melting and electron beam melting machines (1990–2010), as details of these are available in the open literature. Current second generation industrial machines have similar technological underpinnings. Binder jetting is appearing as a viable approach for the realisation of microlattices, and has the advantages of a large choice of materials, of a wide range of scales, of lower costs and of full control over the realisation process.

For SLM and EBM, the formation of the microlattice, parent material micro structure and quality of microlattice structures are dependent on the manufacturing machine optical path, machine control, laser scan strategy, and interface with design software. These will be discussed.

Chapter 4 discusses parent material properties and generic lattice properties. As core materials and energy absorbing devices are the structural applications under discussion, then parent material properties such as modulus, yield stress, ultimate tensile strength, and rupture strain are relevant. In order to focus discussion, three parent materials will be mainly considered, namely stainless steel 316L (SS316L), titanium alloy (Ti 64) and aluminium alloy (AlSi10Mg/AlSi12Mg). Stainless steel 316L is easy to manufacture with, Ti 64 and AlSi10Mg/AlSi12Mg have a good specific properties but are difficult to work with.

Also, to keep the focus in discussion, three lattice topologies are mainly discussed, namely body centered cubic (BCC), body centered cubic with Z vertical struts (BCCZ), and octet truss (OT). The former two microlattices have stable and predictable collapse modes and behaviour. They are also reliable to manufacture using additive manufacturing. The octet truss is more efficient structurally, but is more difficult to manufacture and has more complex progressive collapse modes. These simple geometries can provide benchmarks for more complex (and optimised) topologies.

Generic lattice properties (e.g. in the form of blocks) are stiffness, strength and progressive collapse under compression, tension, shear and multi axial loading. These loadings can take place under static and impact strain rates. For impact, parent material strain rate properties and lattice micro inertia properties are relevant.

Chapter 5 discusses the theoretical structural analysis and synthesis of microlattice architectures. Such activity gives insight into the behaviour of actual microlattices and gives a rigorous basis for realising optimal microlattices. Numerical simulation can use beam elements for micro struts (of use for large scale lattice volumes) or three dimensional modelling for more detailed study of local areas. Finite element analysis is of use for investigating micro inertia and parent material strain rate effects under impact loading. Homogenisation of microlattices is of use for large scale structures. Analytic modelling is useful for parametric modelling (e.g. effect of cell size, strut diameter etc.). Synthesis of optimal microlattices, for a given application, remains problematic however, given the large number of design variables and the complexity of structural response, e.g. impact progressive collapse. However, developing techniques of generative design, voxel based techniques of lattice definition, and general optimisation will be discussed. The latter can include realisation processes. Also, multi functionality is briefly discussed (Xiong et al. 2015), as methods need to be developed to embrace multi functions via formal algorithms.

Chapter 6 introduces the manufacturing processes of photopolymer wave guides and woven metal wire. In the former, either metallic hollow microlattices can be realized using electroless plating or metallic solid microlattices can be realized using investment casting. These processes are in direct competition to selective laser melting and electron beam melting, and the process can push lattice properties into the ultralight performance regime. Associated issues of materials, quality and structural analysis and design are covered. Discussion is extended to shell like structures that can be manufactured using photopolymer wave guides (Shellular). The photo polymer wave guide realisation process can be used for hierarchical materials, and is relevant to the general field of metamaterials. For the woven wire method, conventional large scale structures can be realised, and the procedure can be adapted for shell lattice structures.

Chapter 7 addresses structural design and applications for core materials and energy absorbing devices. As far as core materials are concerned, specific microlattice solutions using selective laser melting, electron beam melting and photopolymer wave guides, and specific materials, are discussed. A critical issue is the skin material for microlattice cores and the design and performance of the core—skin interface is discussed. The potential improvements in beam and panel structural performance such as core stiffness, core failure and localised failure are quantified. As far as impact energy absorption is concerned, the design freedom of cellular topology and to a lesser extent parent material properties, are used to optimise energy absorber behaviour. Structural elements such as lattices, shells and other geometry (e.g. gyroids) will be discussed. Specific performance enhancements over current state of the art are highlighted.

Chapter 8 broadens the discussion and identifies future prospects. As stated earlier in this introduction, the issues discussed here are a subset of architectured cellular materials and metamaterials for a number of structural applications (Schaedler and Carter 2016). Hence, ideas discussed here are related to these wider fields. Specific areas of research and structural engineering science for microlattice core materials and energy absorbing devices are identified. Technology is discussed to give the opportunity for innovative advanced concepts in sandwich cores and energy absorbing devices.

Next, the industrial realisation of microlattice structures is discussed. This includes the full control and quality assurance of the complete realisation process (manufacture, materials, and structure), and certification of the final component for aerospace, bio and other high performance applications (Lloyds Register and TWI 2017). Formalisation of the complete realisation process allows control and optimisation of the process.

A number of suggestions for future structures research are then given covering the topics discussed in the book. Finally, the concept of *design space* and exploring the *design space* are introduced. Specific examples of these design points are discussed in this book, e.g. a specific manufacture, materials and application combinations, and, in this way, detailed understanding of the field can be systematically developed.

To conclude. Metallic microlattice structures, realised using additive manufacture, have significant potential for next generation lightweight structures. Such structures are a subset of architectured cellular materials, and the ideas discussed here are put into the context of this wider field.

References

L.J. Gibson, M.F. Ashby, *Cellular Solids: Structure and Properties*, 2nd edn. (Cambridge University Press, Cambridge UK, 1997)

Q. Li, R.A.W. Mines, R.S. Birch, The crush behaviour of Rohacell 51WF structural foam. Int. J. Sol. Struct. **37**(43), 6321–6341 (2000)

Lloyd's Register, TWI, Guidance notes for certification of metallic parts made by additive manufacturing (2017)

R.A.W. Mines, C.M. Worrall, A.G. Gibson, The static and impact behaviour of polymer composite sandwich beams. Composites **25**(2), 95–110 (1994)

R.A.W. Mines, C.M. Worrall, A.G. Gibson, Low velocity perforation behaviour of polymer composite sandwich panels. Int. J. Imp. Eng. **21**(10), 855–879 (1998)

M.G. Rashed, M. Ashraf, R.A.W. Mines et al., Metallic microlattice materials: a current state of the art on manufacturing, mechanical properties and applications. Mat. Des. **95**, 518–533 (2016)

T.A. Schaedler, W.B. Carter, Architected cellular materials. Ann. Rev. Mat. Res. **46**, 187–210 (2016)

J. Xiong, R.A.W. Mines, R. Ghosh et al., Advanced microlattice materials. Adv. Eng. Mat. **17**, 1253–1264 (2015)

Chapter 2
Some Fundamental Structural Ideas for Conventional Metallic Lattice Structures

Abstract The aim of Chap. 2 is to give a number of fundamental ideas on lattice structures, independent of the use of additive manufacturing technology. These ideas have been developed for a number of years for conventional structures. Aspects covered will be (a) lattice structures as a cellular material, (b) general nomenclature for lattice structures, (c) lattice structures as core materials in sandwich panels, and (d) impact energy absorption in conventional metallic structures. In this way, relevant rigorous engineering science ideas will be identified, and the potential for applying these ideas to additively manufactured microlattice structures will be highlighted.

Keywords Cellular material · Lattice topologies · Sandwich structures · Core materials · Structural impact

2.1 Lattice Structures as a Structural Cellular Material

The classic text on cellular materials is by Gibson and Ashby (1997). The main classes of cellular material discussed in their book are honeycomb and foam. The closest cellular material to the additive manufactured metallic microlattices discussed here is the open cell aluminium foam (Duocel; Ashby et al. 2000). Such foam is investment cast from a polymer precursor. These foams have a variable (stochastic) topology, and cell walls are variable. Open cell foam has the advantages of avoiding trapped air and of ease of analysis (Gibson and Ashby 1997). Detailed analysis of the collapse of cellular materials has been conducted for many years. For example, Zhu et al. (1997) analyse the compression behaviour of open cell polymeric foams to large strains. They develop lattice models for non linear progressive collapse.

The comparison of open cell foams and lattices is discussed by Ashby (2006). The author differentiates between bending dominated structures and stretch dominated structures (see Fig. 2.1). Figure 2.1a shows a pin jointed structure that will collapse as a mechanism. If Fig. 2.1a has welded joints, and the struts bend, it is bending dominated. If a horizontal element is added (in Fig. 2.1b), then for pin jointed structures in Fig. 2.1b, the structure becomes stretch dominated. These aspects can be expressed in terms of Maxwell stability criteria (Ashby 2006). In two dimensions:

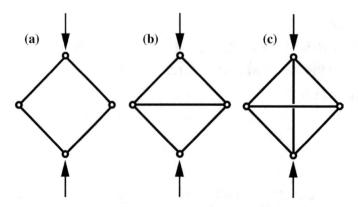

Fig. 2.1 Generic 2D lattice topologies under compression; **a** pin jointed: mechanism, welded joint: member bending, **b** pin joint with cross member: tension dominated, **c** over constrained (Reprinted from Ashby (2006) with permission from the the Royal Society)

$$M = b - 2j + 3 = 0$$

where b is the number of struts, j is the number of frictionless joints. If M is less than zero, the structure is a mechanism (Fig. 2.1a: b = 4, j = 4, M = −1). If M equals zero, the structure is stretched dominated (Fig. 2.1b: b = 5, j = 4, M = 0). If M is greater than zero, the structure becomes self stressed (i.e. internal stresses are generated) as shown with added vertical member in Fig. 2.1c.

In three dimensions, the equation becomes:

$$M = b - 3j + 6 = 0$$

Ashby (2006) states that stretch dominated structures have high structural efficiency and bending dominated structures have low structural efficiency. Foams are usually bending dominated. Ashby (2006) gives a number of three dimensional lattice geometries, and identifies which are stretched dominated and which are bending dominated.

The mechanical properties (stiffness and strength) of stretch dominated lattice structures scale with the relative density in a linear manner whereas for bending dominated structures, the scaling is quadratic. Hence, for a given relative density, the properties for a stretch dominated lattice structure will be higher than for bending dominated lattice structure. However, for the case of stretched dominated, post failure collapse will be more unstable and more complex (involving elastic and plastic buckling), making it less suitable for energy absorption. Ashby (2006) compares lattice structures to honeycombs, foams and woven structures.

A more general, and recent, discussion of these issues is given by Fleck et al. (2010). They introduce the concept of micro architectured materials, and extend Maxwell's rules to include discriminating cell stress and mechanisms, namely in three dimensions:

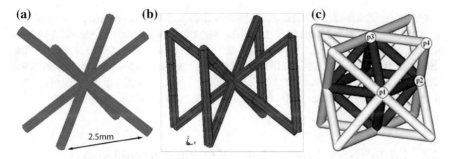

Fig. 2.2 Microlattice topologies; **a** body centred cubic (BCC), **b** BCC with Z strut (BCCZ), and **c** octet truss (OT) (Fig. 2.2c reprinted from Deshpande et al. (2001) with permission from Elsevier)

Table 2.1 3D Maxwell rule for BCC, BCCZ and OT microlattice cells

	BCC	BCCZ	OT
b (number of struts)	8	12	36
j (number of joints)	9	9	14
b − 3j + 6=	−13	−9	0

$$b - 3j + 6 = s - m$$

where s and m count the states of self stress and mechanisms, respectively. A *just rigid* framework has s = m = 0. They also extended discussion to multi scale lattice structures.

The lattice geometries considered in this book are Body Centred Cubic: BCC, BCCZ, and octet truss. These geometries are shown in Fig. 2.2. Table 2.1 gives Maxwell rule values for these. The reasons for selecting the BCC topology in this book is that it is straightforward to manufacture using selective laser melting and electron beam melting (see Chap. 3). Also it has a stable collapse mechanisms and is orthotropic in three dimensions (see Chap. 4). The addition of a Z strut increases stiffness and strength in the given direction. These structures are compared to the octet truss, which is more complex to manufacture (see Chap. 3) and has more complex failure modes (see Chap. 4). However, it is a more efficient structure (see Table 2.1).

2.2   General Nomenclature for Lattice Structures

The above discussion has focused on specific topologies. Zok et al. (2016) give a general discussion on the nomenclature of lattice structures. A BCC lattice (truss) is a cubic system, and is of an elementary family. Other names for BCC structures are octahedral and pyramidal. An octet truss is a cubic system and is of an elementary family, also. Zok et al. (2016) discuss the specific application of sandwich panels, and categorise these as compound trusses. The aim of the work was to present a system for the classification and taxonomy of periodic truss structures. The authors make the point that although simple trusses, e.g. BCC, are not optimal, the performance improves in combination with skins (e.g. sandwich construction). Their work gives a rigorous basis for nomenclature, and for further work on lattice structures.

2.3   Lattice Structures as Core Materials in Sandwich Panels

As far as sandwich structures are concerned, extensive literature occurs on the use of cellular core materials in sandwich structures. Major texts on sandwich structures are by Allen (1969) and by Zenkert (1997). Allen (1969) discusses some core materials in sandwich construction, e.g. honeycomb and balsa wood. The author also identifies the important core properties in sandwich construction. These are shear stiffness and strength, as well as uniaxial stiffness and strength. The authors approach is fully analytic. Zenkert (1997) extends discussion to localised modes and more complex sandwich panel problems.

Abrate (1998) discusses impact on composites sandwich structures, and covers contact laws and impact damage for sandwich construction. Contact panel properties can be adjusted by varying the properties of the cellular core.

Systematic studies of metallic lattice materials used as core materials in sandwich construction have been discussed by Wadley (2006) and Wadley et al. (2003). Wadley (2006) discusses tetrahedral, pyramidal, diamond textile and hollow diamond topology, as well as plated structures. The author uses conventional manufacturing approaches such as stamping, bending and brazing. The author discusses textile and weaving approaches. Materials considered include a wide range of metals. The author compares moduli and strength of the different configurations. The author also considers applications such as thermal management, blast wave mitigation and ballistic (foreign object) impact behaviour. Wadley et al. (2003) discuss metallic lattice structures, manufactured using bending and metal textile approaches, and cores in sandwich beams, with metallic skins. Materials considered include stainless steel, aluminium alloy, and copper alloy. The authors compare compression, and shear responses, and include effects of topology and strain hardening. It should be noted that Dong et al. (2018) extend formed lattice structures into the microlattice scale (1 mm cell size, 100 µm strut diameter). The authors studied stainless steel, and compared compression moduli and strength with other cellular materials.

Dragoni (2013) discusses tetrahedral metallic lattice core beams, and the author systematically varies geometry, e.g. tetrahedral angle, and optimises various mechanical properties. The author uses optimisation techniques to minimise relative density for lattice parameters. From this, it can be seen that lattice cored structures provide the opportunity to optimise core stiffness and strength, and so optimise sandwich panel behaviour.

The progressive collapse of lattice structures include such nonlinear effects as elastic buckling, plastic buckling, plasticity and material rupture. Such failure modes will occur locally in microlattice volumes, and will initiate collapse of lattice structures. The microlattice topology mainly discussed here is body centred cubic, and this fails by plastic bending in a microstrut (see Chap. 4), hence parent material plasticity and rupture are the dominant parameters. In this book, classical plasticity is mainly addressed, however Chap. 4 (Characterisation) discusses scaling effects, as this may start to become significant at scales of the order of magnitude of micro strut diameter. In the numerical simulations in Chap. 5, standard constitutive models in ABAQUS and LS-DYNA are mainly discussed.

As far as material rupture is concerned, material failure will be dependent on the mode of loading and effect of constraint (scale). This has been discussed in general terms by Bao and Wierzbicki (2004). They develop fracture curves in the equivalent strain and stress tri-axiality space, which identifies different failure modes. Ullah et al. (2016) implements three criteria in the finite element package Abaqus for titanium alloy lattices (see Fig. 2.3). Hence localised strut failure in microlattice blocks will be complex, depending on the mode of loading, constraints and ductility of the parent material. Stainless Steel 316L is a ductile material, but Ti 64 and Aluminium Alloy are more brittle.

The other major topology considered in this book is the Octet Truss. This has been discussed earlier, and it is more efficient structurally as compared to BCC,

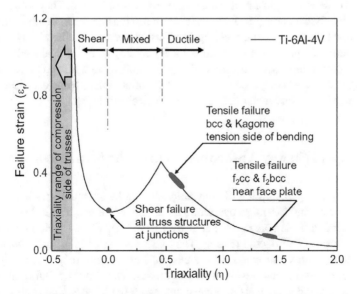

Fig. 2.3 A fracture curve in equivalent strain and tri-axiality space for Ti 64 (Reprinted from Ullah et al. (2016) with permission from Elsevier)

but it suffers more unstable failure modes, viz. elastic and plastic buckling strut collapse. This may initiate instabilities in lattice block collapse. Deshpande et al. (2001) discuss the mechanical properties of the octet truss, in detail. They focus on collapse mechanisms of plastic collapse, anisotropic yield, elastic buckling and effect of geometric imperfections. They create initial collapse surfaces in three dimensional stress space. They conclude that the stiffness and strength values of the octet truss are about half the theoretical maximum values for isotropic voided materials.

Mohr (2005) extends this discussion to generate homogenised models for lattice volumes (see Chap. 5) and compares the models to experimental results (Chap. 7). As far as modelling collapse modes of failures are concerned, the author models finite and small strain plasticity. The author applies the model to lattice three point bend beams. The author highlights the pressure dependence behaviour of octet trusses, which is similar to the behaviour of metal foams (Ashby et al. 2000).

Dong et al. (2015) investigate Ti 64 micro octet lattice structures manufactured using forming and snapped fits (strut section sizes $= 1.588 \times 1.588/0.794$ mm). Cell size is approximately 10 mm. They tested two layer skinned panels under compression, and identified failure modes such as yielding, elastic buckling, and plastic buckling.

The other class of lattice discussed here is with hollow struts. These can be manufactured conventionally (macrostruts) or using photopolymer wave guides (microstruts). The former tend to fail subject to plasticity, and the latter fail due to elastic buckling.

Queheillalt and Wadley (2011) use forming and brazing to construct stainless steel 304L hollow pyramidal lattice structures, with strut diameter $= 3.175$ mm, wall thickness $= 0.13$–0.51 mm and cell size $= 6.40$–12.70 mm. They subject single layer sandwich structures (metals skins) to compression and shear, and they identified failure modes such as plastic yielding, elastic and plastic buckling. They conclude that the dominant parameter controlling local buckling of cylindrical shells is mean tube diameter to tube wall thickness. This gives higher strength and increased post buckling stability as compared to solid struts. Hollow micro struts are discussed in Chaps. 6 and 7, mainly manufactured using electroless plating.

2.4 Impact Energy Absorption in Metallic Structures

The impact energy absorption of conventional metallic structures has been studied for many years, and extensive overviews are given by Lu and Yu (2003) and Jones (2012). Structural elements considered by Lu and Yu (2003) include rings, thin walled members, and ductile tearing. They also discuss cellular materials such as honeycombs, foams and wood. In the case of honeycomb, they discuss dynamic in plane compression and resultant failure modes. The paper by Qui et al. (2009) gives a more detailed analysis of honeycomb and other lattices. The analysis is two dimensional. They consider dynamic compression for honeycomb, rhombus, square, triangular, and Kagome lattices. The impact speeds are in the range of 0–130 ms^{-1}.

For all lattices, the authors show differences in failure modes for impact loading. They use shock wave theory to describe the enhancement to crush stress, σ:

$$\sigma = \sigma_0 + \frac{\rho V_i^2}{\varepsilon_D}$$

where σ_0 is the static crush stress, ρ is the material density, V_i is the impact velocity and ε_D is the densification strain. They show that the average dynamic stresses are higher for the stretch dominated lattices as compared to bending dominated lattices. The analyses are material strain rate independent.

Jones (2012) also discusses the structural impact of conventional structures. The author considers the impact behaviour of structural elements, e.g. beams, plates and shells, from plasticity and buckling points of view. The author includes the effect of material strain rate. Different metals exhibit different strain rate characteristics, but in general, yield stress increases with strain rate. A simple definition of strain rate ($\dot{\varepsilon}$) is:

$$\dot{\varepsilon} = \frac{V_i}{l}$$

where V_i is the impact velocity and l is the gauge length. Hence for a 20 mm solid block impacted at $10\,\mathrm{ms}^{-1}$ (say), the block strain rate is $500\,\mathrm{s}^{-1}$. However, for a lattice block, strain rates will be higher at structural details, due to stress concentration and localised deformation effects.

A simple strain rate constitutive model was suggested by Cowper Symonds:

$$\frac{\sigma_d}{\sigma_s} = 1 + \left(\frac{\dot{\varepsilon}}{D}\right)^{\frac{1}{q}}$$

where D and q are material properties. σ_d is the dynamic stress (yield or UTS), and σ_s is the static stress (yield or UTS). This is applicable to ductile metals and for low to medium strain rates. The model is also attributed to Perzyna (Sasso et al. 2008). Jones (2012) includes strain rate effects in structural analyses, and shows that for a steel clamped beam, the maximum deflection increases by 30% in the impact case due to strain hardening and to strain rate effects.

Alves (2000) discusses material constitutive laws for high strain rates, large strain effects, and for ductile materials. The author concludes that the Cowper Symonds relation does not model high strain rates and large strains accurately, and so further parameters need to be introduced. Sasso et al. (2008) discuss the application of the Johnson Cook model:

$$\sigma = \left(A + B\varepsilon^n\right)\left(1 + C\log\frac{\dot{\varepsilon}}{\dot{\varepsilon}_0}\right)\left(1 - T^{*m}\right)$$

The first set of brackets is for static behaviour, the second set of brackets is for strain rate behaviour and the third set of brackets is for temperature effects. A, B, n, m are material constants, ε is the strain, C is a strain hardening parameter, $\dot{\varepsilon}$ is a strain rate, $\dot{\varepsilon}_0$ is a reference strain rate, and T* is the homologous temperature. The authors make the point that the JC model decouples strain rate, strain and softening effects. The authors compare five different strain rate models and apply them to AISI 1018 CR steel in the strain rate regime of up to 10^3 s^{-1}. The authors compare the accuracy of this model with the Cowper Symonds (Perzyna) model.

Sun and Li (2018) give a wide ranging discussion of the dynamic compression of cellular materials. They cover shock regime and different cellular models. They do not discuss lattice structures but they do discuss open cell metal foams e.g. Duocel (Ashby et al. 2000). The authors differentiate quasi-static/transitional and shockwave regimes. In the former, the internal energy of the whole system is minimised, however cell wall buckling may be influenced by inertia. In the latter, deformations are more localised and hence are not governed by minimal energy.

Hence, for the impact of cellular materials, there are two effects, namely micro inertia and material strain rate dependency. These effects will be dependent on cell topology, parent material and impact velocity. Effects may change lattice failure modes and will change collapse stress. A more detailed discussion related to microlattice structures will be given in Chaps. 5 and 7. Chapter 6 discusses electroless plated (and hence hollow) microlattice structures.

2.5 Conclusions

To conclude: The aim of this chapter has been to give some fundamental structural ideas appropriate to the applications considered in this book. The survey of the literature has necessarily been selective, but there is a large body of rigorous literature covering the behaviour of conventional lattice structures, and many of these ideas can usefully be transferred to additive manufactured microlattice structures and to architectured cellular materials, in general.

It should be noted that large strain plasticity and material rupture behaviour may be dependent on the scale. Hence, these effects may be needed to be included as microlattice structures are driven down in scale.

References

S. Abrate, *Impact on Composites Structures* (Cambridge University Press, Cambridge UK, 1998)
H.G. Allen, *Analysis and Design of Structural Sandwich Panels* (Pergamon Press, Oxford, UK, 1969)
M. Alves, Material constitutive law for large strains and strain rates. J. Eng. Mech. **126**, 215–218 (2000)
M.F. Ashby, The properties of foams and lattices. Phil. Trans. R. Soc. A **364**, 15–30 (2006)

M.F. Ashby, A. Evans, N.A. Fleck et al., *Metal Foams: A Design Guide* (Butterworth Heinemann, Woburn, USA, 2000)

Y. Bao, T. Wierzbicki, A comparative study on various ductile crack formation criteria. J. Eng. Mat. Tech. (ASME) **126**, 314–324 (2004)

V.S. Deshpande, N.A. Fleck, M.F. Ashby, Effective properties of the octet truss lattice materials. J. Mech. Phys. Sol. **49**, 1747–1769 (2001)

L. Dong, V. Deshpande, H.N.G. Wadley, Mechanical response of Ti-6Al-4V octet truss lattice structures. Int. J. Sol. Struct. **60–61**, 107–124 (2015)

L. Dong, W.P. King, M. Raleigh et al., A micro fabrication approach for making metallic mechanical metamaterials. Mat. Des. **160**, 147–168 (2018)

E. Dragoni, Optimal mechanical design of tetrahedral truss cores for sandwich constructions. J. Sand. Struct. Mat. **15**(4), 464–484 (2013)

N.A. Fleck, V.S. Deshpande, M.F. Ashby, Micro architectured materials: past, present and future. Proc. R. Soc. A **466**, 2495–2516 (2010)

L.J. Gibson, M.F. Ashby, *Cellular Solids: Structure and Properties*, 2nd edn. (Cambridge University Press, Cambridge UK, 1997)

N. Jones, *Structural Impact*, 2nd edn. (Cambridge University Press, Cambridge, UK, 2012)

G. Lu, T. Yu, Energy Absorption of Structures and Materials (CRC Press (Woodhead Publishing Limited), Cambridge, UK, 2003)

D. Mohr, Mechanism based multi surface plasticity model for ideal truss lattice material. Int. J. Sol. Struct. **42**, 3235–3260 (2005)

D.T. Queheillalt, H.N.G. Wadley, Hollow pyramidal lattice truss structures. Int. J. Mat. Res. **102**(4), 389–400 (2011)

X.M. Qui, J. Zhang, T.X. Yu, Collapse of periodic planar lattices under uniaxial compression, part II: dynamic crushing based on finite element simulation. Int. J. Imp. Eng. **36**, 1231–1241 (2009)

M. Sasso, G. Newaz, D. Amodio, Material characterization at high strain rate by Hopkinson bar tests and finite element optimisation. Mat. Sci. Eng. A **487**, 289–300 (2008)

Y. Sun, Q.M. Li, Dynamic compressive behaviour of cellular materials: a review of phenomena, mechanism and modelling. Int. J Imp. Eng. **112**, 74–115 (2018)

I. Ullah, M. Brandt, S. Feih, Failure and energy absorption characteristics of advanced 3D truss core structures. Mat. Des. **92**, 937–948 (2016)

H.N.G. Wadley, Multi functional periodic cellular metals. Phil. Trans. R. Soc. A **364**, 31–68 (2006)

H.N.G. Wadley, N.A. Fleck, A.G. Evans, Fabrication and structural performance of periodic cellular metal sandwich structures. Comp. Sci. Tech. **63**, 2331–2343 (2003)

D. Zenkert, *An Introduction to Sandwich Construction* (EMAS Limited, Cradley Heath, UK, 1997)

H.X. Zhu, N.J. Mills, J.F. Knott, Analysis of the high strain compression of open cell foams. J. Mech. Phys. Sol. **45**(11/12), 1875–1899 (1997)

F.W. Zok, R.M. Latture, M.R. Begley, Periodic truss structures. J. Mech. Phys. Sol. **96**, 184–203 (2016)

Chapter 3
Additive Manufacturing Processes and Materials for Metallic Microlattice Structures Using Selective Laser Melting, Electron Beam Melting and Binder Jetting

Abstract The additive manufacturing processes discussed here have been selected for their significance for the selected structural applications, e.g. core materials and energy absorbing materials. Selective laser melting and electron beam melting are mature (industrial) processes, whereas binder jetting (and associated techniques) is currently under intense development. As far as selective laser melting and electron beam melting are concerned, the controlling parameter is the beam scanning strategy, which defines the dimensions and quality of the microlattice. Also, the parent material will influence the realisation process, the final quality of the microlattice and structural performance. In this discussion, three main materials will be discussed: namely, stainless steel 316L, titanium alloy Ti 64, and aluminium alloy AlSi10/12Mg. Stainless steel 316L is widely discussed in the literature, and Ti 64 and AlSi10/12Mg are lower density but more highly reactive materials.

Keywords Selective laser melting · Electron beam melting · Binder jetting · Laser scanning · Additive manufacture materials · Metallic glasses

3.1 Selective Laser Melting (SLM)

Selective laser melting is an additive manufacturing process, in which fine metallic powder is selectively melted using a laser to form intricate structures. The process was developed by the Fraunhofer Institute in the 1990s (Meiners et al. 1999), and then exploited in the 1990s and 2000s by small scale spin off companies, such as SLM Solutions and MCP in Germany, and MTT in the United Kingdom. These will be called 'first generation machines'. They were, in the main, developed in academic institutions and small companies, and hence they are well documented and are relatively simple. Around 2010, the selective laser melting process made it into mainstream industrial production, as exemplified by the takeover of MTT by the UK Company Renishaw in 2010, who have developed the technology further. This means that 'second generation machines' are more closed (due to intellectual property rights), are more complex (to address quality issues) but they use the basic processes from 'first generation machines'. The technology continues to develop.

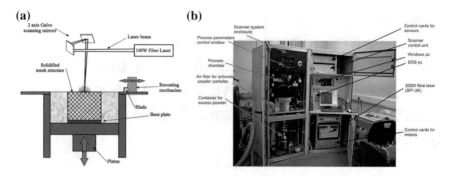

Fig. 3.1 a Schematic of the SLM process showing the laser optical path. **b** Photo of MCP Realizer II SLM250 machine used at the University of Liverpool (Photo taken in 2010)

The approach in this book is to introduce the reader to the 'first generation' selective laser melting machines as described by Rehme (2010) and Tsopanos et al. (2010), in order to define the basic technology and the influence of manufacturing process and materials used on lattice configurations. The development of the 'first generation' machines tended to be empirical in nature, in which machines were designed and then adjusted for optimal manufacturing conditions (Mullen et al. 2008). It should be noted that the selective laser melting machines are general purpose, and so are not optimised for microlattice structures. For a general discussion of additive manufacture, where selective laser melting and electron beam melting are puts into the context of other additive manufacturing processes, see the Gibson et al. (2015) for all additive manufacture processes and Milewski (2017) for additive manufacture of metals only. A useful and practical overview of additive manufacturing processes is given by Redwood et al. (2017), and an up to date overview is given by Thompson et al. (2016).

Figure 3.1a gives a schematic of the SLM process. The basic processes is as follows. A layer of metallic atomised powder (grain size of the order of 40 μm) is spread onto a steel bed plate. The laser is then fired, with a controlled laser path and laser exposure time. The powder is selectively laser melted. A new layer of powder is then evenly spread over the area, and the laser is again operated. In this way, a complex geometry is created layer by layer. The process is slow, taking about 30 s per layer (depending on the complexity of the job). Given that each layer is the order of 50 μm, a typical structural component of 20 mm dimension has over 1000 layers, which means the manufacturing of components can be a number of hours. Figure 3.1b gives a photo of the actual selective laser melting machine used at Liverpool from 2003 to 2010.

As the manufacturing time increases so the potential for degradation of the optical path during manufacture becomes an issue—which is a motivation for including monitoring and feedback control in the manufacturing process. Another issue is processing time (with associated accuracy issues). The movement of the laser at the build plane is dependent on the inertia of the moving parts in the optical path, which

Fig. 3.2 SS316L stainless steel microlattice blocks and single struts on MCP realiser II SLM machine bed plate. Note different strut angles and block supports

defines parameters such as overshoot and settling time. Figure 3.2 shows microlattices and micro struts manufactured using an MCP Realiser II SLM 250 machine in 2009 (Tsopanos et al. 2010). Figure 3.3 shows the detail from a microlattice node for titanium alloy (Mines et al. 2013). The single spot laser scan strategy was used, as indicated by the black spots in Fig. 3.3.

The thermodynamics of these processes are complex, and are only now being addressed in detail. Suffice to say that the final lattice quality depends on the optical path (laser, lens, galvanometers), as well as on the laser power and exposure time, the laser scanning strategy and speed of manufacture. SLM manufacturing machines will continue to improve. One example of this is the selective laser melting machine made by Xact Metal. This machine is based on a standard selective laser melting machine layout, but the galvanometers that control laser mirrors are restricted to moving around two axes (not three). This reduces the cost of the machine by 50%, and allows the adjustment of laser spot size during manufacture (Molitch-Hou 2017).

It should be noted that many structural research papers on microlattice structures neglect giving details of the manufacture (including laser parameters), which detract from the contributions.

The focus of this book is on a load bearing performance of these microlattice structures, and Chap. 4 addresses the materials characterisations issues. However, this chapter continues with a detailed discussion of selective laser melting machine control and other AM processes.

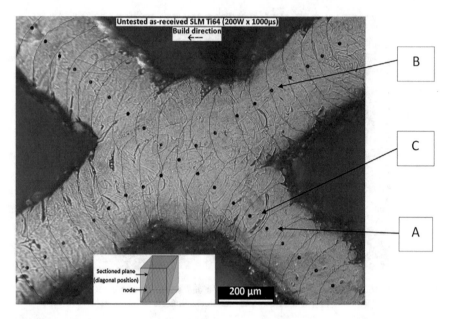

Fig. 3.3 Detail of Ti 64 microlattice BCC block node, manufactured using single spot laser. Black dots are notional foci for laser. Laser emits at point A, moves on to point B, moves on to other structures on the bed plate, and then returns to point C at the next layer. The section was taken at an angle in the block

3.2 SLM Laser Scan Strategy and Microstrut Quality

The conversion of CAD files to selective laser melting instructions is not covered here, given that there are many proprietary systems available and that they are machine dependent. Also, this is a fast changing technology. However, there is scope to select process accuracy and speed. As microlattices become more complex and optimised, including changing laser parameters and even materials during the process, so this conversion software becomes more complex.

The main parameter defining scale and accuracy of the microlattice is the laser scan strategy. This can be (a) single spot, (b) contour, (c) hatch and (d) combination of contour and hatch.

For the author's microlattice structures, the paper by Mullen et al. (2008) described the method used to realise the microlattice structures. Special laboratory based software (Manipulator) was written to create wire meshes for the BCC geometry. These were then sliced and converted to machine format (MCP Realiser II SLM 250). The single spot laser scan strategy was used.

Ghouse et al. (2017) compared single spot and contour hatch laser scan strategies (see Fig. 3.4). The contour hatch strategy improves the quality of the microstrut and allows near horizontal struts. However, feature size is larger than the single spot approach, and structures are more time consuming to manufacture.

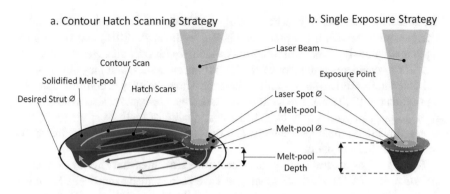

Fig. 3.4 Laser scan strategies: **a** contour hatch, **b** single exposure [reprinted from Ghouse et al. (2017) under Creative Commons Attributions License (CCBV)]

Messner (2017) suggested an improved 2D sizing method, using graph based storage for struts. An algorithm was then developed to directly slice the members. The author increased the manufacturing speed by 100 and computer memory used by one tenth. This approach is applicable to a wide variety of additive manufactured models.

Tsopanos et al. (2010) discussed the effects of varying laser power and exposure time on the dimensions and quality of the final microlattice. Changing laser parameters will change strut diameter, and so these parameters can be used to control lattice density. The authors proposed that the laser process is characterised by:

$$E_p = LP \times LX$$

where E_p is the laser energy delivered to the powder bed, LP is the laser power and LX is the laser exposure time. The advantages of the single spot approach are speed and minimum strut diameter. Disadvantages are variation in strut surface and loss of quality of struts, especially near the horizontal due to stepping (Hasan 2013).

Ghouse et al. (2017) developed the single spot approach further and systematically studied loss of quality for low strut angles. They used a pulsed laser to improve micro powder melting and so solidification. The authors also characterised the process in terms of specific enthalpy (ΔH):

$$\Delta H = \frac{AP}{\rho\sqrt{\pi Du\varphi^3}}$$

where A = absorptivity, P = laser power, ρ = powder density, D = thermal diffusivity, u = laser speed, φ = laser spot diameter. They considered stainless steel 316L and titanium alloy Ti 64. They showed good correlation between ΔH and strut thickness, and they suggested ΔH is a better characterising parameter compared to laser energy (E_p). The authors also noted that, due to the complexity of actual microlattices, microlattice generation will move from solid to lined models, and so a single spot approach is the preferred choice. Vrana et al. (2018) conducted an in depth quality analysis for AlSi10Mg microlattices with strut diameter down to 500 μm. They considered a number of laser scanning strategies, and they also realised and tested hollow micro struts (wall width approximately 300 μm and strut diameter approximately 1200 μm). The authors concluded that the contour laser scan strategy gives good quality micro struts, that the best quality of inclined struts (at 35°) is achieved using minimal laser energy, and the authors gave the laser conditions for optimal micro strut quality.

Given the loss of strut quality for microlattice structures that have low angles to the horizontal, BCC and BCCZ topologies have been studied extensively. However, Chap. 2 discussed the octet truss, and identified the higher structural efficiency for this microlattice. However, this topology has struts at or near the horizontal for all orientations. Hence, in order to realize an octet truss the contour and hatch laser scan approaches are required. Tancogne-Dejean et al. (2016) used a contour hatch laser scan strategy to realise stainless steel 316L octet truss with a minimum diameter of 534 μm. This is larger than the 200 μm for single spot approach (Tsopanos et al. 2010).

Quantification of the mechanical performance of microlattice structures is dependent on dimensional accuracy and surface quality. This includes circularity, diameter variation and surface quality. Hasan et al. (2011) discussed this for a single spot laser scan strategy for titanium alloy Ti 64. Microlattice quality can be optimised using manufacturing parameters, but post SLM treatment can also be used. For example, Pyka et al. (2012) used chemical etching and electrochemical polishing to improve the surface of titanium alloy Ti 64 microlattice. Full definition of the microlattice geometry is required for data reduction in property measurement (Chap. 4) and for finite element modelling (Chap. 5).

From the previous discussions, it can be seen that the manufacture of selective laser melted microlattices is complex and subject to many parameters. If selective laser melted microlattices are to be used in industrially significant applications, then the full realisation process needs to be defined, controlled and optimised. Papers are starting to appear in the literature on this. For example, Sing et al. (2018) quantified the effects of laser power, laser scan speed and layer thickness using regression analysis and analysis of variance. A more general, design for additive manufacture approach was discussed by McMillan et al. (2017). In this, the authors used a finite difference computational approach to systematically vary process parameters. Their model needs to be calibrated using specific cases e.g. an inclined cylinder.

To conclude this section, the researcher in selective laser melting microlattice structures needs to put the realisation of their microlattice structures into the context of the complete realisation process. This is discussed in more detail in Chap. 8.

3.3 Electron Beam Melting (EBM) Process

The Electron Beam Melting process is discussed by Gibson et al. (2015), and is a direct competitor to selective laser melting. The process uses a high energy electron beam to induce fusion between metal powders. Gibson et al. (2015) summarised the differences between selective laser melting and electron beam melting processes. The minimum feature size, resolution and surface finish of an electron beam melting process is typically larger than for a selective laser melting process. However, the process is faster, suitable for porous structures and other complex geometries. Smith et al. (2016) discussed the electron beam melting process in detail, and used an ARCAM A1 EBM machine with titanium alloy Ti 64. They discussed how to optimise the realisation process for rhomboid microlattice cells. De Formanoir et al. (2016) took titanium alloy octet truss microlattices manufactured using the ARCAM A1 EBM machine and chemically etched them to improve surface quality. Strut diameters from 770 to 1660 μm were studied. They concluded that chemical etching reduces dimensional variations. Suard et al. (2015) studied Ti 64 1000 μm single struts produced by electron beam melting. The authors quantified porosity, shape and size, and surface roughness. They developed a general methodology to predict the quality of additively manufactured microlattice structures.

Hence, to conclude this section, electron beam melting has certain advantages over selective laser melting, although micro strut diameters are larger than those produced using selective laser melting.

3.4 Materials used in Selective Laser Melting and Electron Beam Melting Processes

One of the advantages of selective laser melting and electron beam melting is that they can use conventional structural materials, e.g. stainless steel 316L, titanium alloy Ti 64 and aluminium alloy AlSi10/12Mg. The processes use atomised powder derived from conventional materials. Selective laser melting and electron beam melting machine manufacturers define these materials, e.g. Table 3.1. These materials need to be fully defined and studied in order to adhere to realisation quality assurance processes (see Chap. 8).

Polmear (2006) discussed the conventional processing, microstructure and application of Ti 64 and AlSi10/12Mg. As far as Ti 64 is concerned, it contains both alpha and beta phases, which give relatively high tensile strengths and improved formability. As far as AlSi10/12Mg are concerned, the addition of magnesium improves age hardening, and doubles yield strength. These alloys also have good corrosion resistance.

Papers are appearing in the literature to improve manufacturability and mechanical physical properties. For example, Vrancken et al. (2014) took titanium alloy Ti 64 powder and mixed in molybdenum powder (10%). The molybdenum powder was regular with size of five microns, whereas the titanium alloy Ti 64 powder was spherical of the order of 10–30 μm. The molybdenum suppressed micro structure transformation from beta phase to alpha martensite. The effect on mechanical properties is to increase rupture strain from 7.3 to 20.1%. This means that lattice blocks become more stable in collapse (see Chap. 4). Wei et al. (2015) added nanohydrosynatite powder (size 200 μm) to stainless steel 316L powder. The authors showed that the optimal content was 5%, and that for this, elastic modulus increased by 25%. Other mechanical properties were similar. It should be noted that alloy additions are relevant to multi functionality, e.g. bio compatibility, wear, heat transfer etc. (see Chap. 8).

Li et al. (2018) added alumina to selective laser melted stainless steel 316L. This was achieved by physical mixing of the powders. From miniature tensile tests (1.5 by 3 mm section) it was shown that an increase of approximately 10% in yield stress and approximately 5% increase in ultimate stress occurred for one per cent by weight alumina. The authors also conducted compression tests on FCCZ microlattices, and showed a 30% increase in specific strength and a 15% in specific energy absorption for one per cent by weight of alumina.

It should be noted that the selective laser melting process can be adjusted to optimise parent material micro structure, and hence mechanical properties for existing materials. For the case of stainless steel 316L, by adjusting laser parameters Liu et al. (2017) increased yield by a factor of three, the ultimate tensile strength by 10% and the rupture strain by 20% over wrought annealed materials, for 3 mm diameter tensile specimens. Wang et al. (2018) used a similar approach and material to increase yield stress by 100%, ultimate tensile strength by 20% and rupture strain by 100% over wrought steel. These issues are of relevance for fully optimising the complete realisation process (see Chap. 8). Hence, it can be concluded that the topology and structural design is not the only method for improving structural performance.

Table 3.1 Range of materials for SLM [from Renishaw], EBM [from ARCAM] and conventional BJ [from EXONE]. Materials discussed in this book in *italic*

SLM	EBM	Conventional BJ
In718-0405	*Ti6Al4V*	*SS316L*
CoCr-0404	*Ti6Al4VELI*	SS17-PH
AlSi10Mg-0403	Titanium grade 2	SS304L
SS316L-0410/407	Cobalt chrome	SS 420+Bronze
Ti6Al4V ELI-0406	Alloy 718	SS316+Bronze
In625-0402		
M300 maraging steel		

Sources
www.renishaw.com/en/data-sheets-additive-manufacturing-17862 (Accessed 5 November 2018)
www.arcam.com/technology/products/metal-powders/ (Accessed 5 November 2018)
www.exone.com/resources/materials (Accessed 5 November 2018)

Care is needed in the terminology for powder bed fusion processes. A case in point is direct metal laser sintering. Originally this process involved sintering, but more recently partial and full melting of the powder occurs, depending on the material. Kruth et al. (2005) discussed four binding mechanisms, namely solid state sintering, chemically induced binding, liquid phase sintering (partial melting) and full melting. As an example of direct metal laser sintering for microlattice structures, Crupi et al. (2017) realised Ti 64 BCC microlattice structures using an EOSINT—M280 DMLS system. In this, a 170 W laser was used to melt globules of material (in a similar manner to SLM). Cell size was 2–2.5 mm, strut diameters were 0.3–0.5 mm. Similar block compression results to Mines et al. (2013) were achieved. Hence it can be concluded that the performance and method of each powder bed fusion machine needs to be considered in detail for a given application. Just to complicate matters further, the additive manufacturing machine company Concept Laser calls their process 'Laser Cusing' (Concept Laser 2018).

3.5 Binder Jetting (BJ) Approach

One of the currently most active areas of additive manufacture of metals is 'binder jetting'. Again, terminology is confusing, so two specific cases will be discussed here, namely binder jetting of microlattice structures and binder jetting of nano lattice structures. In the former process, a fusion deposition method using metal particles in a binder is used to create a three dimensional geometry (Bose et al. 2018, Jakus et al. 2015). This technology is based on metal injection moulding (MIM), which is a well established technology with a wide selection of materials. Hence, the process is cheaper and faster as compared to selective laser melting and electron beam melting. The metal powder and binder are fusion deposition modelled (FDM) into the defined geometry. The green part is then passed through binder removal, and then sintered in an oven. Some shrinkage (of up to 25%) occurs, and the part is 99% consolidated. The Desktop Metal company quotes a resolution of 50 μm (Desktop Metal 2018). The process allows the production of closed cell cellular and porous structures.

Vangapally et al. (2017) used an EXONE binder jetting machine and binder jet process for stainless steel 316L. Their processed structure has rectangular cells, with wall thickness of 1 mm, with similar cell size. They encountered shrinkage of up to 4%, and so this has to be accounted for in the green part design. Tang et al. (2016) used an EXONE M Lab 3DP machine to create SS316L rectangular unit cells (strut diameter of 500 μm) in the form of a cylindrical specimen [13 mm diameter (4 cells) and 25 mm length (10 cells)].

In general, struts size and dimensional accuracy is currently inferior to selective laser melting and electron beam melting processes. Also, parent material integrity and microstructure is inferior to selective laser melting and electron beam melting. It should be noted that an alternative method of binding jetting is to lay the powder down first and then to impregnate the particles with resin. The green part is then sintered. This process is the basis for new HP metal printers (Hewlett Packard 2018a). Papers

Table 3.2 Process selection for SLM, EBM, and conventional BJ [from Gokuldoss et al. (2017)]

	SLM	EBM	Conventional BJ
Material type	Ti/Al difficult	Not Al/Mg/Zn based	Not amorphous alloys
Available technology	Mature	Not fully certified for process parameters	Mature (moulding)
Post processing	Not required	Not required	Large amount needed
Accuracy	Restricted by powder	Restricted by powder	Wide range powders
	Good accuracy	Good accuracy	Significant shrinkage
Lead time/cost	High	High	Faster/lower cost
Properties	Tune using processing	Less controllable	Inferior properties
Surface quality	Restricted powder size	Restricted powder size	Wide range powder size

have yet to appear in the literature that characterise the microlattice produced using this process. The process may not be able to be used to create closed cell cellular structures.

Gokuldoss et al. (2017) compared selective laser melting, electron beam melting and binder jetting. Features of binder jetting are removal of binders, suitability for support structures, fast, cheap (due to material), wide range of materials, reduced mechanical properties, geometric tolerance (due to shrinkage). Hence, the process needs to be further developed for intricate lattice structures.

Table 3.2 compares the SLM, EBM and BJ processes.

However, unlike the selective laser melting and electron beam melting processes, the binder jetting process can be scaled in size, and the process has been used to realise sub microlattice structures with silver nano particles, using an Aerosol Jet Printer (AJ300) (Saleh et al. 2017). They used the point wise printing strategy, and created 1.5 mm cubed blocks with 250 mm cell size and approximately 50 μm strut diameter. Blocks were tested under compression loading.

Another binder jetting manufacturing technique (known as drop cast) has recently been published for reduced scale of metallic lattices (nickel) down to nano scale (Vyatskikh et al. 2018). In this process, an ultraviolet curable metal based photo resist was dropped cast using a Photonic Professional GT Nanoscribe two photon polymerization 3D printer. The samples were then pyrolysed to volatilise the organic constituents. The results were nickel octet nano lattices with a unit cell size of two microns and a strut diameter of 300–400 nm. Cubic blocks ($4 \times 4 \times 4$ cells, $40 \times 40 \times 40 \,\mu$m) were tested under compression, and it was shown that specific strength was competitive with other additive manufacturing processes. From the supplementary information for this paper, failure modes were strut buckling followed by rupture. Detailed modelling of these failures will require scaled constitutive relations as discussed previously (Chap. 2). It should be noted that the technique has good throughput and that the maximum block size is 3 mm cubed (300 cell cubed) depending on the

Nanoscribe machine. Also other organometallics can be used. Hence structural ideas developed for microlattices can be extended to nano lattices. Application of such metallic nano lattices could include MEMS, micro batteries, micro bots and micro medical procedures (Vyatskikh et al. 2018).

There is still the scale gap with strut diameter from 100–0.5 μm (Vyatskikh et al. 2018). It is proposed that the binder jet methods will be developed for this scale gap in the future.

An alternative general term for all of the above processes is 'Bind and Sinter' (Bracket 2018). Also the term 'material jetting' is sometimes used (Gibson et al. 2015).

Hence, the use of the binder jet process allows microlattices to be constructed with a wider variety materials at cheaper cost, and allows reduction in scale to nano scale. However, this book focuses mostly on selective laser melting and electron beam melting as these currently give the highest quality microlattice structure.

3.6 Amorphous Metals (Metallic Glasses)

Amorphous metals (metallic glasses) lack the crystallinity of normal metals, which gives them distinct properties. Ashby and Greer (2006) discuss these differences. In general, amorphous metals are brittle and lack plasticity. However they become more ductile at small scale. Brothers and Dunand (2005) discuss ductility at length scales of 25 μm, and Huang et al. (2007) show increased ductility in going from length scales of 6 mm, to 3 mm, and to 1 mm.

Amorphous metals (metallic glass) can now be made using additive manufacturing (specifically selective laser melting), and the small section size of additive manufactured cellular materials and the potential cost control of additive manufacturing formation processes can be exploited to enhance ductility (Pauly et al. 2013). These authors construct a 3D scaffold structure. Currently amorphous metals (metallic glasses) at small scale are manufactured using thermo plastic forming (Kumar et al. 2011). The improvements in ductility of in-plane loaded metallic glass honeycomb is discussed by Liu et al. (2016). It should be noted that metallic glasses can be electroless plated on photopolymer wave guides, and this is discussed in Chap. 6. Chapter 5 discusses the nonlinear optimisation of a microlattice metallic glass. Hence such materials have the potential to exploit enhanced plasticity at low length scale.

3.7 Additive Manufacture in Metals Using Multiple Materials

Additive manufacturing with multiple materials has been pursued over the last few years (Vaezi et al. 2018). Commercial systems are now available for polymeric materials, e.g. HP Jet Fusion (Hewlett Packard 2018b). Multiple material additive manufacturing is less well developed for metals (Vaezi et al. 2018). The authors identify issues such as: contamination, bonding, data processing, process interruption and materials development. As far as the additive manufacturing processes discussed in this book are concerned, binder jetting and material jetting (with sintering) provide the most potential routes for multiple materials (Bandyopadhyay and Heer 2018). Some preliminary discussions of the effect of having regions in additive manufactured materials with different properties is discussed by Hiller and Lipson (2009). The above issues are discussed further in Chap. 5 in relation to voxels.

It should also be noted that Gümrük et al. (2018) took their selectively laser melted stainless steel 316L BCC microlattice, and electroless plated them with nickel phosphorus alloy. Such coating enhances structural performance (see Chap. 7), and gives potential for multi functionality, e.g. wear and heat transfer.

3.8 Conclusions

To conclude: three manufacturing processes have been discussed here, namely selective laser melting, electron beam melting and binder jetting. The selection of process for a given application depends on a number of factors including material selection. Table 3.2 lists some of these factors for selective laser melting, electron beam melting and binder jetting. Other additive manufactured metal processes are available, as discussed by Milewski (2017). From the structures researcher point of view, the process has to be selected and fully defined and validated before characterising (Chap. 4), simulating (Chap. 5) and structural testing (Chap. 7).

From the materials point of view, existing materials can be optimised, additions to materials can be considered or new classes of materials (e.g. metallic glasses or multiple materials) can be considered.

The next topic of discussion is the characterisation of microlattice structures using the micro strut tensile test and compression tests of microlattice blocks. The selective laser melting process is the focus of discussion. Both static and impact loading regimes are covered.

References

M.F. Ashby, A.L. Greer, Metallic glasses as structural materials. Scr. Mater. **54**, 321–326 (2006)

A. Bandyopadhyay, B. Heer, Additive manufacturing of multi material structures. Mat. Sci. Eng. R **129**, 1–16 (2018)

A. Bose, C.A. Schuh, J.C. Tobia et al., Traditional and additive manufacturing of a new tungsten heavy alloy alternative. Int. J. Refractory Mat. Hard Mat. **73**, 22–28 (2018)

D. Bracket, Binder and sinter additive manufacturing flexibility (2018), www.tctmagazine.com/blogs/guest-column/binder-and-sinter-additive-manufacturing-flexibility/ (Accessed 18th November 2018)

A.H. Brothers, D.C. Dunand, Ductile bulk metallic glass foams. Adv. Mater. **17**(4), 484–486 (2005)

Concept Laser (2018), www.concept-laser.de/technologie.html (Accessed 19th November 2018)

V. Crupi, E. Kara, G. Epasto et al., Static behavior of lattice structures produced via direct metal laser sintering technology. Mat. Des. **135**, 246–256 (2017)

C. De Formanoir, M. Suard, R. Dendievel et al., Improving the mechanical efficiency of electron beam melted titanium lattice structures by chemical etching. Add. Manuf. **11**, 71–76 (2016)

Desktop Metal (2018), www.desktopmetal.com/products/studio/ (Accessed 19th November 2018)

S. Ghouse, S. Babu, R.J. Van Arkel et al., The influence of laser parameters and scanning strategies on the mechanical properties of stochastic porous materials. Mat. Des. **131**, 498–508 (2017)

I. Gibson, D. Rosen, B. Stucker, *Additive Manufacturing Technologies: 3D Printing, Rapid Prototyping and Direct Digital Manufacturing*, 2nd edn. (Springer, New York, USA, 2015)

P.K. Gokuldoss, S. Kollar, J. Eckert, Additive manufacturing processes: selective laser melting, electron beam melting and binder jetting—selection guidelines. Materials **10**(6), 672–692 (2017)

R. Gümrük, A. Usun, R. Mines, The enhancement of the mechanical performance of stainless steel microlattice structures using electroless plated nickel coatings. MDPI Proc. **2**(8), 494 (2018)

R. Hasan, Progressive collapse of titanium alloy microlattice structures manufactured using selective laser melting. Ph.D. Thesis, University of Liverpool, 2013

R. Hasan, R. Mines, P. Fox, Characterisation of selectively laser melted TI 6Al 4V microlattice struts. Procedia Eng. **10**, 536–541 (2011)

Hewlett Packard (2018a), www8.hp.com/us/printers/3d-printers/metals.html (Accessed 19th November 2018)

Hewlett Packard (2018b), www8.hp.com/us/printers/3d-printers/metals.html (Accessed 19th November 2018)

J. Hiller, H. Lipson, Design and analysis of digital materials for physical 3D voxel printing. Rap. Proto. J. **15**(2), 137–149 (2009)

Y.J. Huang, J. Shen, J.F. Sun, Bulk metallic glasses: smaller is softer. Appl. Phys. Lett. **90**, 081919-1–3 (2007)

A.E. Jakus, S.L. Taylor, N.R. Geisendorfer et al., Metallic architectures from 3D printed powder fused liquid inks. Adv. Funct. Mat. **25**(45), 1–10 (2015)

J.P. Kruth, P. Mercelis, J. Van Vaerenbergh et al., Binding mechanisms in selective laser sintering and selective laser melting. Rap. Prot. J. **11**(1), 26–36 (2005)

G. Kumar, A. Desai, J. Schroers, Bulk metallic glass: the smaller the better. Adv. Mat. **23**, 461–476 (2011)

X. Li, H.J. Willy, S. Cheng et al., Selective laser melting of stainless steel and alumina composite: experimental and simulation studies on processing parameters, microstructure and mechanical properties. Mat. Des. **145**, 1–10 (2018)

L. Liu, Q. Ding, Y. Zhong et al., Dislocation network in additive manufactured steel breaks strength—ductility trade off. Mat. Today **21**(4), 354–361 (2017)

Z. Liu, W. Chen, J. Carstensen et al., 3D metallic glass structures. Acta Mater. **105**, 35–43 (2016)

M. McMillan, M. Leary, M. Brandt, Computationally efficient finite difference method for metallic additive manufacturing: a reduced order DFAM tool applied to SLM. Mat. Des. **132**, 226–243 (2017)

W. Meiners, C. Over, K. Wissenbach, et al., Direct generation of metal parts and tools by selective laser powder re-melting (SLPR), in *Proceedings of Solid Freeform Fabrication Symposium*, Austin, Texas, USA (1999)

M.C. Messner, A fast efficient direct sizing method for slender member structures. Add. Manuf. **18**, 213–220 (2017)

J.O. Milewski, *Additive Manufacturing of Metals: From Fundamental Technology to Rocket Nozzles, Medical Implants and Custom Jewelry (Springer Series in Materials Science 258)* (Springer, New York, USA, 2017)

R.A.W. Mines, S. Tsopanos, Y. Shen et al., Drop weight impact behaviour of sandwich panels with metallic microlattice cores. Int. J. Imp. Eng. **60**, 120–132 (2013)

M. Molich-Hou, What exactly makes Xact Metals's metal 3D printing so cheap? (2017), www.engineering.com/3DPrinting/3DPrintingAricles/ArticleID/15300 (Accessed July 2018)

L. Mullen, R.C. Stamp, W.K. Brooks, et al., Selective laser melting: a regular unit cell approach for the manufacture of porous, titanium, bone in growth constructs, suitable for orthopedic applications. J. Biomat. Mat. Res. **89**(B), 325–334

S. Pauly, L. Lober, R. Petters et al., Processing metallic glasses by selective laser melting. Mat. Today **16**(1/2), 37–41 (2013)

I. Polmear, *Light Alloys: From Traditional Alloys to Nano Crystals*, 4th edn. (Butterworth and Heinemann, Oxford, UK, 2006)

G. Pyka, A. Burakowski, G. Kerckhofs et al., Surface modification of Ti 6Al 4V open porous structures produced by additive manufacturing. Adv. Eng. Mat. **14**(6), 363–370 (2012)

B. Redwood, F. Schoffer, B. Garret, The 3D Printing Handbook: Technology, Design and Applications. 3D Hubs BV, Netherlands (2017)

O. Rehme, *Cellular Design for Laser Freeform Fabrication* (Cuvillier Verlag, Gottingen, Germany, 2010)

M.S. Saleh, C. Hu, R. Panat, Three dimensional micro architected materials and devices using nano particle assembly by pointwise spatial printing. Sci. Adv. **3**, e1601986 (2017)

S.L. Sing, F.E. Wiria, W.Y. Yeong, Selective laser melting of lattice structures: a statistical approach to manufacturability and mechanical behavior. Robot. Comput. Integr. Manuf. **49**, 170–180 (2018)

C.J. Smith, F. Derguti, E. Hernandez Nava et al., Dimensional accuracy of electron beam melting (EBM) additive manufacture with regard to weight optimized truss structures. J. Mat. Proc. Tech. **229**, 128–138 (2016)

M. Suard, G. Martin, P. Lhuissier et al., Mechanical equivalent diameter of single struts for the stiffness prediction of lattice structures produced by electron beam melting. Add. Manuf. **8**, 124–131 (2015)

T. Tancogne Dejean, A.B. Spierings, D. Mohr, Additively-manufactured microlattice materials for high specific energy absorption under static and dynamic loading. Acta Mater. **116**, 14–28 (2016)

Y. Tang, Y. Zhou, T. Hoff et al., Elastic modulus of 316L stainless steel lattice structures fabricated by binder jetting process. Mat. Sci. Tech. **32**(7), 648–656 (2016)

M.K. Thompson, G. Moroni, T. Vaneker et al., Design for additive manufacturing: trends, opportunities, considerations and constraints. CIRP Ann. **65**(2), 737–760 (2016)

S. Tsopanos, R.A.W. Mines, S. McKown, et al., The influence of processing parameters on the mechanical properties of selectively laser melted stainless steel microlattice structures. J. Manuf. Sci. Eng. (ASME) **132**, 041011-1–12 (2010)

M. Vaezi, S. Chianrabutra, B. Mellor et al., Multiple material additive manufacturing—Part 1, A review. Virt. Phys. Protot. **8**(1), 19–50 (2018)

S. Vangapally, K. Argarwal, A. Sheldon et al., Effect of lattice design and process parameters on dimensional and mechanical properties of binder jet additively manufactured stainless steel 316 bone scaffolds. Procedia Manuf. **10**, 750–759 (2017)

R. Vrana, D. Koutny, D. Palousek et al., Selective laser melting strategy for fabrication of thin struts usable in lattice structures. Materials (MDPI) **11**, 1763 (2018)

B. Vrancken, L. Thijs, J.P. Kruth et al., Microstructure and mechanical properties of a novel β titanium metallic composite by selective laser melting. Acta Mater. **68**, 150–158 (2014)

A. Vyatskikh, S. Delalande, A. Kudo et al., Additive manufacturing of 3D nano architected metals. Nat. Commun. **9**(593), 1–8 (2018)

Y.M. Wang, T. Voisin, J.T. McKeown et al., Additively-manufactured hierarchical stainless steel with high strength and ductility. Nat. Mat. **17**, 63–71 (2018)

Q. Wei, S. Li, C. Han, W. Li et al., Selective laser melting of stainless steel/nano hydroxyapatite composites for medical applications: microstructure, element distribution, crack and mechanical properties. J. Mat. Proc. Tech. **222**, 444–453 (2015)

Chapter 4
Parent Materials and Lattice Characterisation for Metallic Microlattice Structures

Abstract The aim of this chapter is to describe the measurement of parent material properties, mostly using micro strut tensile tests, and to discuss the failure modes of microlattice blocks of selected topologies, mostly under compressive loading. A number of microlattice parent materials (SS316L, Ti 64, AlSi10/12Mg), topologies (BCC, BCCZ, OT), lattice block loadings (compression, tension, shear, combined impact) will be discussed. Modes of failure of microstruts will be discussed and specific stiffness and strength properties compared. Both static and impact loading will be discussed. Parent material behaviour includes large plastic strains and material rupture.

Keywords Tensile testing · Material plasticity · Material rupture · Dynamic loading · Strain rate · Lattice block stiffness · Lattice block strength · Modes of failure

4.1 Micro Strut Tensile Tests (Static)

Material properties can be measured using standard material tests or scaled down to actual microstrut size. ISO 17296-3:2014 lists the standards needed to measure various mechanical properties in additive manufacture (British Standards 2014). In this book, we are mostly interested in tensile strength (ISO 6891-1 British Standards 2016). Such a standard defines specimen dimensions and testing procedures. The TWI—Lloyds Register guidance notes for certification (Lloyds Register—TWI 2017) refers directly to the ISO 17296-3 (and hence ISO 6891-1) for certification purposes. Seifi et al. (2016) discussed the need for additive manufacturing material qualification, from the point of the view of the USA. This is all part of the complete control of the additive manufacture realisation process. However, there is a need for some reduced size tensile tests to reflect scale effects in manufacturing, build quality, and plasticity and rupture. This is the approach taken here.

Lavvafi et al. (2014) studied the bending and tension failure of stainless steel 316 LVM wires with diameters of 100–154 μm. In the tensile tests, 30 mm gauge length specimens were glued to tabs, and tested in a screwdriven Instron testing machine.

© The Author(s), under exclusive licence to Springer Nature Switzerland AG 2019   33
R. Mines, *Metallic Microlattice Structures*, SpringerBriefs
in Structural Mechanics, https://doi.org/10.1007/978-3-030-15232-1_4

Tensile fracture of the stainless steel 316LVM specimens showed a dimpled fracture and a high reduction in area at failure, typical of stainless steel. They derived results similar to standard stainless steel.

Gümrük and Mines (2013) developed similar tension tests for stainless steel 316L SLM Microlattice struts manufactured with parameters discussed in Tsopanos et al. (2010). Figure 4.1 shows details of the test. In this, individual struts were adhesively bonded on to tabs, and a clip gauge was adhesively bonded on to the specimen, and the setup was mounted into an Instron servohydraulic machine. Figure 4.2 gives stress strain data for a laser power of 140 W and exposure time of the 1000 μs. Laser parameters affect the quality of the specimen and were discussed by Tsopanos et al. (2010) and by Shen et al. (2010). Also, in deriving stress strain data, the cross-sectional area of the microstrut has to be assessed. Shen (2009) and Hasan (2013) discussed microstrut quality for stainless steel and Ti 64, respectively. Table 4.1 compares derived data with other data from the literature. Mines et al. (2013) gave stress strain data for Ti 64 microstruts. Generally, the quality of the Ti 64 specimen is not as good as stainless steel 316L and derived properties compare unfavourably with text book data. There was slight contamination of Ti 64 microlattice, due to the use of the selective laser melting machine for multiple powders (Hasan et al. 2011; Hasan 2013). Brandao et al. (2017) discussed cross contamination issues, which need to be addressed by certification methodologies.

Apart from the quality of the microstrut specimen, the other two issues are scaling of plasticity and mechanisms of rupture. Microstrut specimens tend to be 200–300 μm in diameter.

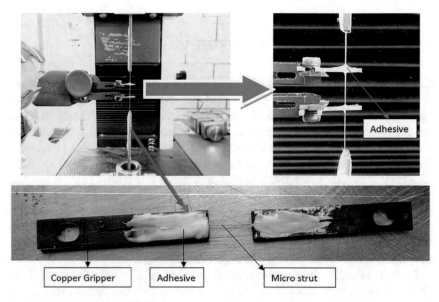

Fig. 4.1 Single strut test for SS316L manufactured using MCP Realizer II SLM250. Clip gauge adhesively bonded to microstrut specimen

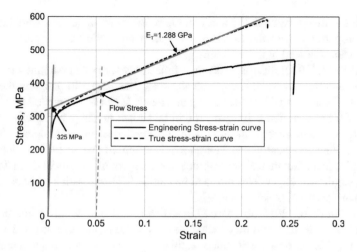

Fig. 4.2 Quasi-static stress strain data for a SS316L micro strut. Laser power = 140 W, Exposure time = 1000 μs

Table 4.1 Comparison of mechanical property data from additive manufacture (AM) with textbook (TB) values [*Mines et al. (2013) and **Leary et al. (2016)]

	SS316L*			Ti 64*			AlSi12Mg**		
	AM	TB	AM/TB	AM	TB	AM/TB	AM	TB	AM/TB
E (GPa)	140	205	0.67	45	115	0.39	–	–	–
$\sigma_{0.2}$ (MPa)	144	310	0.46	240	898	0.27	236	460	0.51
σ_{UTS} (MPa)	280	620	0.45	280	996	0.28	434	391	1.11
ε_f	0.16	0.50	0.32	0.01	0.10	0.10	0.05	0.09	0.56

Gümrük and Mines (2013) did not study the microstructure of SLM stainless steel 316L parent material in detail. Fairly obviously, the formation of selective laser melted micro struts is a complex process: micro powder is melted, is solidified and is tempered (as the laser spot moves away from the build area). Luo and Zhao (2018) surveyed thermal analysis methods in the powder fusion additive manufacturing process and developed a finite element model for the process. Wang et al. (2018) and Liu et al. (2018) studied, in detail, selective laser melted stainless steel 316L microstructure, and the optimisation of laser parameters for optimised properties and microstrut. Wang et al. (2018) gave a detailed micro structural analysis of stainless steel 316L selectively laser melted vertical struts in the diameter range of 5 mm down to 250 μm. The authors do not give details of their laser scan strategy. They conduct tension and micro hardness tests. They conclude that strut diameter influences microstructure, and hence strength and ductility. Materials properties improve with strut diameter.

It should therefore be noted that there is scope to improve microstructure (and hence properties) from the experimentally measured microlattice data from the authors values quoted in this chapter.

Hutchinson (2000) discussed scaling of plasticity for polycrystalline metals and identifies scaling effects in tension for the scale less than 100 μm, and the effect of torsion for a scale of less than 170 μm. Hence, it can be concluded that scaling effects will begin at about 100 μm. It will be shown in Chap. 5 that concentrated plasticity occurs in material during microlattice collapse, and so scale may become an issue at stress concentrations. However, it is proposed that given the variability in the additive manufacture realisation process, scaling for microlattices discussed here is not a significant effect. This needs to be clarified. Most plasticity models used in Chap. 5 are classical plasticity models.

Bültmann et al. (2015) discussed the scalability of microstruts in the diameter range of 300–740 μm, and they include tests with multiple microstruts. The motivations for this were to use standard testing machines and to mitigate handling of specimen issues. The number of microstruts varied from 7 to 37. The struts where realised using SLM and cross hatch laser scan approach. The authors compared yield, ultimate tensile strength and elongation to rupture for stainless steel 316L, for different strut diameters, for a number of struts, for different strut lengths, for different relative surface areas and angles of strut. They concluded that selective laser melting microstruts give different properties to standard tensile tests, and that the measured mechanical properties are similar for all strut diameters.

One class of materials that is scale dependent is metallic glasses (amorphous metal alloys). These were introduced in Chap. 3. Kumar et al. (2011) and Huang et al. (2007) discuss this issue. Metallic glasses can now be 3D printed (Pauly et al. 2013). The optimisation of these materials is discussed by Carstensen et al. (2015), and will be dealt with in Chap. 5.

The other material property data that needs to be addressed is that of rupture. Carlton et al. (2016) discussed damage evolution and failure mechanisms in additive manufactured stainless steel. They used standard specimens (diameter = 6.35 mm), and quantified pores in tested specimens and damage evolution. They identified different failure mechanisms depending on pores, and depending on realisation processes.

Bao and Wierzbicki (2004) discussed various ductile crack formation criteria depending on triaxiality. This work was discussed in Chap. 2. Ullah et al. (2014) applied these criteria to additive manufacture Ti 64 Kagome microlattices (diameter = 0.6 and 1.2 mm). The rupture strain criteria (ε_f) are:

$$\varepsilon_f = \frac{D_1}{(1+3\eta)} + D_2 \quad -\tfrac{1}{3} < \eta \leq 0 \ \text{for shear failure}$$
$$\varepsilon_f = D_3\eta^2 + D_4\eta + D_5 \quad 0 < \eta \leq \eta_T \ \text{for mixed mode}$$
$$\varepsilon_f = D_6 + D_7 e^{(-D_6\eta)} \quad \eta_T \leq \eta \ \text{for ductile failure}$$

where η represents stress triaxiality, and η_T is the transition point. D_i values are material properties.

Figure 2.3 (Chap. 2) gave a graphical representation of this. Hence material rupture depends on scale (triaxiality) and on the mode of loading. Ullah et al. (2014) calibrated the damage evolution model in Abaqus to capture this (see Chap. 5). Good agreement between numerical and experimental results for lattice block shear and compression were shown.

Wang and Li (2018) conducted single strut tensile tests for selectively laser melted Ti 64 with diameters of 300–1200 μm. They characterised, in detail, the variations in diameter of the struts, and expressed their measurements in terms of Feret diameters (which characterises section shape factor). The authors studied in detail the strut tensile rupture behaviour, and they fit constitutive and failure behaviour to the Johnson Cook constitutive model (as discussed in Chap. 2). They applied the data to a finite element analysis model of compressed BCC lattice blocks.

4.2 Micro Strut Tensile Tests (Dynamic)

Chapter 2 discussed conventional metals strain rate behaviour and some constitutive relations that capture strain rate, large strains and temperature effects. Such data is relevant for the analysis of dynamic progressive collapse of microlattice blocks. Gümrük et al. (2018) extend their microstrut tensile testing methodology to dynamic tests in the parent material strain rate regime of low rate ($8.3 \times 10^{-3}\ \mathrm{s}^{-1}$), medium rate ($5.8$–$8.1\ \mathrm{s}^{-1}$) and high rate ($2050$–$6610\ \mathrm{s}^{-1}$) for stainless steel 316L. The intermediate regime tests used a single shot Instron E3000 machine, and the high rate regime tests used a multi specimen split Hopkinson bar. Figure 4.3 compares data, with data measured for conventional specimens by Langdon and Schleyer (2004). From this, it can be concluded that measured data is consistent with standard data, and Table 4.2 gives the Cowper Symonds coefficients for strain rate modelling. More sophisticated strain rate models are available (see Chap. 2) but it is felt that the simple Cowper Symonds (Perzyna) model is appropriate at this time. As yet, material rupture behaviour of microstruts at high strain rates has not been studied.

Austin et al. (2017) test Ti 64 statically under tension and dynamically under compression, and in the latter a split Hopkinson bar was used. Their specimens were standard size and specimens were manufactured using electron beam melting and laser metal deposition. In the latter process, the metal powder is sprayed on to the build area and melted using a laser. They fit their experimental results to the Johnson Cook constitutive model, and they found an increase in dynamic yield stress of about 40%.

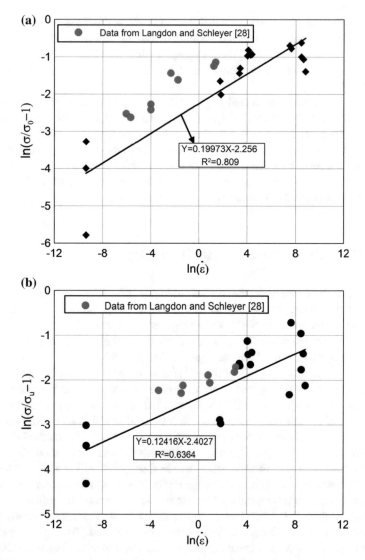

Fig. 4.3 Comparison of dynamic materials data from microstrut tensile tests for SS316L for **a** yield stress (σ_0) and **b** ultimate stress (σ_u) between micro struts and conventional test data. Reference [28] refers to Langdon and Schleyer (2004)

4.3   Microlattice Block Characterisation (Static and Dynamic)

The next characterisation step is to study simple microlattice blocks under various loading conditions, but mainly in compression. A number of parameters can be identified, namely:

Table 4.2 Cowper Symonds coefficients (D, q) for strain rate modelling (Gümrük et al. 2018)

Source	D (s^{-1})	q	Comment
Langdon and Schleyer (2004)	429–2721	4.1–5.8	–
Burgan (2001)	240	4.74	–
Experiment (yield stress)	4852	4.08	Without split Hopkinson bar data
Experiment (yield stress)	80,737	5.01	All data
Experiment (UTS)	252 × 10^6	8.05	All data
Suggested	17 × 10^6	12	Estimation

1. Effect of material: stainless steel 316L, Ti 64, AlSi10/12Mg
2. Effect a cell scale: cell size = 1.25, 2.5 mm
3. Effect of cell topology: body centred cubic (BCC), body centered cubic with Z strut (BCC, Z), octet truss (OT)
4. Effect of manufacturing process: selective laser melting, electron beam melting, direct metal laser sintering, binder jetting
5. Effect of mode of loading: compression, tension, shear, combined
6. Effect of loading rate: low, intermediate, high.

It should be noted that testing cell blocks should take into account existing testing standards, e.g. ISO BS 13314:2011 (British Standards 2011). Table 4.3 highlights 8 papers used in the subsequent discussion and Table 4.4 summarises the specific stiffness, strength and failure modes.

To set the scene, the paper by Gümrük and Mines (2013) is discussed (stainless steel 316L, BCC, static, SLM, compression). Figure 4.4 shows block specimens for cell size of 1.25 and 2.5 mm, with failure modes under static loading. For the 2.5 mm cell size case, collapse occurs with plastic hinges near the nodes and cell collapse is fairly well distributed through the lattice block. The collapse mode in stainless steel 316L BCC is stable due to the ductility of the parent material, and bending dominated collapse behaviour occurs. In their paper, the cell size is reduced from

Table 4.3 Lattice block compression data for parameter comparison (main parameter change in *italic*)

Source	Material	Cell size (mm)	Topology	Process	Loading[a]	Rate[b]	Constraint[c]
Gümrük and Mines (2013)	SS316L	1.25/2.5	BCC	SLM	C	S	UC
Gümrük et al. (2013)	SS316L	1.25/1.5 2.0/2.5	*BCC, F2BCC, BCCZ*	SLM	C,T,S	S	C/UC

(continued)

Table 4.3 (continued)

Source	Material	Cell size (mm)	Topology	Process	Loading[a]	Rate[b]	Constraint[c]
Gümrük et al. (2018)	SS316L	2.5	BCC	SLM	C	*D*	UC
Mines et al. (2013)	*Ti 64*	2.5	BCC	SLM	C	S	UC
Leary et al. (2016)	*AlSi12Mg*	7.5	BCC, BCCZ, FCC FCCZ, FBCCZ	SLM	C	S	UC
Tancogne-Dejean et al. (2016)	SS316L	3.1	*OT*	SLM	C	S/D	UC
Ozdemir et al. (2016)	Ti 64	5	BCC	*EBM*	C	S	UC
Harris et al. (2017)	SS316L	2	'BCCZ'	SLM	C	*D*	UC

[a]*C* Compression, *T* Tension, *S* Shear
[b]*S* Static, *D* Dynamic
[c]*C* Constrained, *UC* Unconstrained

2.5 to 1.25 mm in four increments. The distribution of failed cells changes from continuous (2.5 mm) to crush shear bands (1.25 mm). Block stiffness and strength increases in a continuous manner with strut aspect ratio.

Gümrük et al. (2013) extended their discussion to tension, shear and multi axial block loading and to BCCZ, F2BCC topology (stainless steel 316L, static, SLM). They also differentiate between unconstrained block (free to laterally deform) and constrained block (at block top and bottom). It should be noted that body centred cubic is sensitive to lateral constraints, and so failure modes will be dependent on lateral constraint. As far as mode of loading is concerned (stainless steel 316L, 2.5 mm cell size, BCC), for the case of shear, struts fail by bending at nodes and shear collapse occurs across the block. For the case of block tension, struts fail by tension and bending giving rise to hourglass block behaviour. The plastic collapse envelopes for cell sizes 1.25 and 2.5 mm are shown in Fig. 4.5.

As far as the compression of BCC, Z is concerned, introduction of Z struts initiates buckling behaviour, which initiates shear bands in the block. It should be noted that Gümrük et al. (2013) adjust selective laser melting manufacturing parameters for the Z struts, and measure a significant increase in specific energy absorption. As far

Table 4.4 Comparison of lattice block compressive specific stiffness, specific strength and failure mode for selected block parameters (Density SS316L: 7860 kg m^{-3}, Ti 64 4680 kg m^{-3}, Al Alloy 2700 kg m^{-3})

Source	Topology	Cell size (mm)	Number cells	Strut diameter (μm)	Block stiffness 10^6 Pa (specific stiffness)(10^3 Nm kg^{-1})	Block strength 10^6 Pa (specific strength) (Nm kg^{-1})	Block density kg m^{-3}	Block relative density	Block fail mode
Gümrük and Mines (2013) (SS316L)	BCC	2.5	8 × 8 × 8	200	19 (65.3)	0.5 (1718)	291	0.037	Continuous
Gümrük et al. (2013) (SS316L)	BCCZ	2.5	8 × 8 × 8	200	420 (1404)	1.48 (4950)	299	0.038	Shear band
	F2BCC	2.5	8 × 8 × 8	200	150 (354)	1.48 (3491)	424	0.054	Shear Band
Gümrük et al. (2018) (SS316L)	BCC Dyn.	2.5	4 × 4 × 4	200	28 (96.2)	0.65 (2234)	291	0.037	Continuous
Mines et al. (2013) (Ti 64)	BCC	2.5	8 × 8 × 8	263	25 (89.0)	2 (7117)	281	0.060	Shear band
Leary et al. (2016) (AlSi12Mg)	BCC	7.5	10 × 10 × 10	1000	150 (652)	4 (17391)	230	0.085	Continuous
	BCCZ	7.5	10 × 10 × 10	1000	600 (2290)	9 (34,351)	262	0.097	Shear band
Tancogne-Dejean et al. (2016) (SS316L)	OT (Static)	3.1	7 × 7 × 7	534	2250 (1060)	45 (21,206)	2122	0.27	Planar front
	OT (Dyn)	3.1	7 × 7 × 7	534	3410 (1607)	58 (27,332)	2122	0.27	Planar front

(continued)

Table 4.4 (continued)

Source	Topology	Cell size (mm)	Number cells	Strut diameter (μm)	Block stiffness 10^6 Pa (specific stiffness)(10^3 Nm kg^{-1})	Block strength 10^6 Pa (specific strength) (Nm kg^{-1})	Block density kg m^{-3}	Block relative density	Block fail mode
Ozdemir et al. (2016) (Ti 64)	Static/BCC	5	$5 \times 5 \times 5$	1000	389 (−607)	20 (31,201)	641	0.137	Planar front
	Dyn/BCC	5	$5 \times 5 \times 5$	1000	(−)	35(54602)	641	0.137	Planar front
Harris et al. (2017) (SS316L)	Static/F2CCZ	2	Cylinder	400	5000 (3262)	50 (32,615)	1533	0.195	Planar front
	Dyn/F2CCZ (100 m/s)	2	Cylinder	400	500 (326)	75 (48,924)	1533	0.195	Continuous

Fig. 4.4 Experimentally measured and finite element simulation deformation for BCC SS316L blocks at various block compression strains. **a** Cell size = 2.5 mm, **b** cell size = 1.25 mm

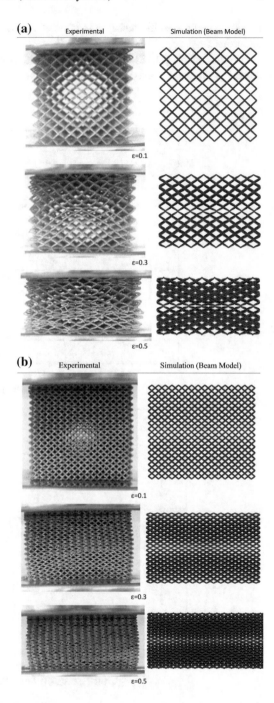

as compression specific strength is concerned the addition of a Z strut maximises performance.

Gümrük et al. (2018) extended their discussion to block impact loading (low velocity $\lambda_l = 7$ ms^{-1}) and to material strain rate effects. The impact effects for struts

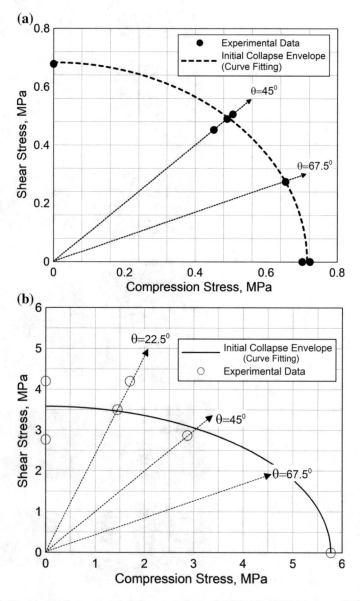

Fig. 4.5 Experimentally measured collapse envelopes for SS316L BCC blocks. **a** Cell size = 2.5 mm, **b** cell size = 1.25 mm

have been discussed previously. As far as the block (stainless steel 316L, BCC, compression) is concerned, increasing impact velocity increases crush stress, but keeps the impact failure mode similar to the static case, e.g. plastic hinges in struts near the nodes. It is proposed that the increase in crush stress is due to material strain rate effects, and this is discussed further in Chap. 5 on finite element modelling, and Chap. 7 on structural applications.

As far as changing material parent material is concerned, Mines et al. (2013) discussed Ti 64 blocks (BCC, cell size 2.5 mm, compression). The main features of Ti 64 compared to stainless steel 316L are low density (60%) and lower rupture to failure (20%). An issue with Ti 64 is hot isostatic processing (HIP), to improve properties. As built Ti 64 blocks fail along a band of cells. The microstruts fail due to large strain and rupture in the vicinity of the nodes, giving rise to a smooth stress-strain collapse curve. The latter can be improved by HIPing. Replacing stainless steel 316L by Ti 64 for the body centred cubic case increases specific stiffness by a factor of 33%. The failure mode changes from continuous to shear band. The slight contamination in selective laser melted Ti 64 was discussed previously, and this may affect block collapse behaviour and properties.

As far as changing the parent material to aluminium alloy is concerned, Leary et al. 2016 tested AlSi10Mg SLM microlattice, and they test BCC, BCCZ, FCC Z and FBCCZ topologies. As far as BCC (cell size = 0.5 mm, strut diameter equals 1 mm) is concerned, block collapse was similar to stainless steel 316L, with plastic hinges at the struts near the nodes and distributed block cell wall collapse. The ductility of AlSi10Mg is in between stainless steel 316L and Ti 64, and so block collapse is dominated by parent material plasticity. As far as BCCZ is concerned, failure is associated with Z strut collapse and BCC microstrut bending. Block crush characteristics change to shear band for the BCCZ case. Replacing stainless steel 316L by AlSi10/12Mg increases specific stiffness by a factor of 10, and strength by a factor of 10, also.

Tancogne-Dejean et al. (2016) tested blocks with octet truss topology (stainless steel 316L, SLM, cell size = 3.08 mm, strut diameter = 0.2–0.74 mm). For relative density = 0.27 (diameter equals 0.5 mm) collapse behaviour is stable, and micro strut failure is dominated by bending and block collapse occurs throughout the block, but initiates at block edges. Going from BCC to OT increases specific strength by 12.5, and changes the failure mode from continuous to planar front. Tancogne-Dejean et al. (2016) repeated tests for impact loading (velocity of impact = 20 ms^{-1}). Slight increases in block crush stress occur, and cell failure occurs at one end of the specimen and progresses through the block. The increase in block crush stress is attributed to material strain rate effects.

Ozdemir et al. (2016) discussed electron beam melting manufactured microlattice blocks (Ti 64, cell size = 5 mm, diameter = 1/1.414 mm, diamond structure (similar to BCC), block compression). In the static loading case, block failure progressed from the lower loaded face. Initial failure of the block was unstable giving a large drop off in load. Thereafter, failure was less oscillatory. Failure modes are different to those of Ti 64 BCC in Mines et al. (2013) although block crush stress data are similar. Changing manufacturing process from SLM to EBM (for BCC Ti 64) and increasing strut diameter from 263 to 1000 μm, increased specific strength by a factor of 18.4. Ozdemir et al. (2016) also conducted impact tests (velocity of impact = 178 ms^{-1}), and showed similar failure modes to the static case. For the EBM case, the failure load changed but the failure mode stayed the same, indicating material strain rate effects.

Harris et al. (2017) tested SLM, stainless steel 316L, BCCZ cylindrical blocks (cell size = 2 mm, diameter equals 0.2–0.4 mm) under high velocity impact loading (V_i = 150 ms^{-1}). They showed that failure modes were similar to the static case, and that the increase in crush stress was due to material strain rate effects.

From the above, it can be concluded that there is a complicated relationship between the six parameters under consideration. Only a specific set of parameters have been considered here, but considered papers do discuss other parameters. There is a need to develop systematic failure maps, for the different parameters, to give a rigorous basis to the selection of parameters. Realisation process and microstrut quality issues also need to be included.

Although progressive collapse of lattice structures in this book is the focus, it should be noted that structural performance will be influenced by residual stresses and fatigue performance. Mercelis and Kruth (2006) have investigated residual stresses for selective laser melted stainless steel 316L solid specimens, and showed changes in the stress of the order of 30%. Van Hooreweder et al. (2017) discussed the fatigue behaviour of selective laser melted titanium alloy Ti 64 microlattices. They investigated the effects of three post build treatments, namely stress relieving, hot isostatic processing and chemical etching. They concluded that a combination of hot isostatic processing and chemical etching is the best.

Ghouse et al. (2018) discussed the fatigue behaviour of Ti 64 alloy microlattices produced by selective laser melting, with various laser scanning strategies including the single spot approach. They identified notch sensitivity, ductility, internal porosity and material microstructure as important parameters. They found an eight per cent increase in fatigue strength, for optimised laser parameters.

4.4   Conclusions

This chapter has been an overview on characterising microlattice materials and blocks. The restricted range of materials and topology from selective laser melting and electron beam melting have been discussed. The wide range of microlattice structural behaviours have been identified for static, impact and multi axial loading. These behaviours are dependent on parent material properties and the quality of the manufacture.

This data is discussed further in Chap. 5, where theoretical and numerical models are discussed. Chapter 7 gives further case studies specific to the structural applications under discussion.

References

D.C. Austin, M.A. Bevan, A.D. East, et al., Microstructural investigation and impact testing of additively-manufactured Ti-6Al-4V in Characterisation of Minerals, Metals and Materials, in *The Minerals, Metals and Materials Society*, ed. by S. Ikhmayies, et al. (2017)

Y. Bao, T. Wierzbicki, A comparative study on various ductile crack formation criteria. J. Eng. Mat. Tech. (ASME) **126**, 314–324 (2004)

A.D. Brandao, R. Gerard, J. Gumpinger et al., Challenges in additive manufacturing of space parts: powder feedstock cross contamination and its impact on end products. Materials (MDPI) **10**, 522 (2017)

British Standards (2014): BS EN ISO 17296—3: 2014. Additive manufacturing—general principles. Part 3: Main characteristics and corresponding test methods. British Standards Institution

British Standards, BS ISO 13314: 2011. Mechanical testing of metals—ductility testing—compression tests for porous and cellular metals. British Standards Institution. UK (2011)

British Standards (2016): ISO 6892-1: Metallic materials—tensile testing—Part 1: method of test at room temperature. British Standards Institution

J. Bültmann, S. Merkt, C. Hammer, et al., Scalability of the mechanical properties of selective laser melting produced micro struts. J. Laser Appl. **27**(S2) S29206-1–7 (2015)

B. Burgan, Elevated temperature and high strain rate properties of offshore steel. Construction Institute. Offshore Technology Report. 2001/20 (2001)

H.D. Carlton, A. Haboub, G.F. Gallegos, et al., Damage evolution and failure mechanism in additively-manufactured stainless steel, Mat. Sci. Eng. **A651**, 406–414 (2016)

J.V. Carstensen, R. Lotfi, J.K. Guest, et al., Topology optimisation of cellular materials with maximised energy absorption, in *Proceedings of ASME 2015 International Design Engineering Technical Conferences and Computers and Information in Engineering Conference IDETC/CIE*, Boston, Mass., USA, 2–5 Aug 2015 (2015)

S. Ghouse, S. Babu, K. Nai et al., The influence of laser parameters, scanning strategies and material on the fatigue strength of a stochastic porous structure. Add. Manuf. **22**, 290–301 (2018)

R. Gümrük, R.A.W. Mines, S. Karadeniz, Static mechanical behaviours of stainless steel microlattice structures under different loading conditions. Mat. Sci. Eng. **A586**, 392–406 (2013)

R. Gümrük, R.A.W. Mines, S. Karadeniz, Determination of strain-rate sensitivity of microstruts manufactured using the selective laser melting method. J. Mat. Eng. Perf. (ASM) **27**(3), 1016–1032 (2018)

R. Gümrük, R.A.W. Mines, Compressive behaviour of stainless steel microlattice structures. Int. J. Mech. Sci. **68**, 125–139 (2013)

J.A. Harris, R.E. Winter, G.J. McShane, Impact response of additively manufactured metallic hybrid lattice materials. Int. J. Imp. Eng. **104**, 117–191 (2017)

R. Hasan, Progressive collapse of titanium alloy microlattice structures manufactured using selective laser melting, Ph.D. Thesis, University of Liverpool, 2013

R. Hasan, R. Mines, P. Fox, Characterisation of selectively laser melted Ti-6Al-4V microlattice struts. Procedia Eng. **10**, 536–541 (2011)

Y.J. Huang, J. Shen, J.F. Sun, Bulk metallic glasses: smaller is softer. Appl. Phys. Lett. **90**, 081918-1–3 (2007)

Hutchinson JW (2000) Plasticity at the micron scale, Int.J.of Sol. and Struct., 37:225-238

G. Kumar, A. Desai, J. Schroers, Bulk metallic glass: the smaller the better. Adv. Mat. **23**, 461–476 (2011)

G.S. Langdon, G.K. Schleyer, Unusual strain rate sensitive behaviour of AISI 316 L austenitic stainless steel. J. Strain. Anal. **39**(1), 71–86 (2004)

H. Lavvafi, J.R. Lewandowski, J.J. Lewandowski, Flex bending fatigue testing of wires, foils and ribbons, Mat. Sci. Eng. A **601**, 123–130 (2014)

M. Leary, M. Mazur, J. Elambasseril et al., Selective laser melting (SLM) of AlSi12 Mg lattice structures. Mat. Des. **98**, 344–357 (2016)

L. Liu, Q. Ding, Y. Zhong et al., Dislocation network in additive manufactured steel breaks strength ductility trade off. Mat. Today **21**(4), 354–361 (2018)

Lloyds Register/TWI, Guidance notes for certification of metallic parts made by additive manufacturing, March 2017. LR/TWI, London (2017)

Z. Luo, Y. Zhao, A survey of finite element analysis of temperature and thermal stress fields in powder bed fusion additive manufacturing. Add. Manuf. **21**, 318–332 (2018)

P. Mercelis, J.P. Kruth, Residual stresses in selective laser sintering and selective laser melting. Rapid Prototyp. J. **12**(5), 254–265 (2006)

R.A.W. Mines, S. Tsopanos, Y. Shen et al., Drop weight impact behaviour of sandwich panels with metallic microlattices cores. Int. J. Imp. Eng. **60**, 120–132 (2013)

Z. Ozdemir, E. Hernandez Nava, A. Tyas et al., Energy absorption in lattice structures in dynamics: experiments. Int. J. Imp. Eng. **89**, 49–61 (2016)

S. Pauly, L. Lober, R. Petters et al., Processing metallic glasses by selective laser melting. Mat. Today **16**(1/2), 37–41 (2013)

M. Seifi, A. Salem, J. Beuth et al., Overview of materials qualification needs for metal additive manufacturing. J. Mat. **68**(3), 747–764 (2016)

Y. Shen, High performance sandwich structures based on novel metal cores, Ph.D. Thesis, University of Liverpool, 2009

Y. Shen, S. McKown, S. Tsopanos et al., The mechanical properties of sandwich structures based on metal lattice architectures. J. Sand. Struct. Mat. **12**, 159–180 (2010)

D. Tancogne-Dejean, A.B. Spierings, D. Mohr, Additively manufactured metallic microlattice materials for high specific energy absorption under static and dynamic loading. Acta. Mater. **116**, 14–28 (2016)

S. Tsopanos, R.A.W. Mines, S. McKown, et al., The influence of processing parameters on the mechanical properties of selective laser melted stainless steel microlattice structures. J. Manuf. Sci. Eng. **132**, 041011-1–12

I. Ullah, J. Elambasseril, M. Brandt et al., Performance of bio inspired Kagome truss core structures under compression and shear loading. Comp. Struct. **118**, 294–302 (2014)

B. Van Hooreweder, Y. Apers, K. Lietaert et al., Improving the fatigue performance of porous metallic bio materials produced by selective laser melting. Acta Biomater. **47**, 193–202 (2017)

X. Wang, J.A. Muniz Lerma, O. Sanchez Mata et al., Microstructure and mechanical properties of stainless steel 316L vertical struts manufactured by laser powder bed fusion process. Mat. Sci. Eng. A **736**, 27–40 (2018)

Z. Wang, P. Li, Characterisation and constitutive model of tensile properties of selective laser melted Ti6Al4V struts for microlattice structures. Mat. Sci. Eng. A **725**, 350–358 (2018)

Chapter 5
Theory, Simulation, Analysis and Synthesis for Metallic Microlattice Structures

Abstract The focus for applications in this book are core materials in sandwich beams and panels, and energy absorbing devices. Hence, microlattice structures will not only be subject to elastic deformation, but also to plastic deformation, buckling and rupture. Also, these responses may take place under impact loading. Thus, theoretical models developed will have to address these non linear and transient effects. The simplest lattice finite element model is modelling the strut as a set of beams. Such an approach is appropriate for complete modelling of large scale microlattice structures. For more detailed modelling, the selected number of cells can be modelled using three dimensional solid elements. Such models discriminate three dimensional plasticity, material rupture and the interaction between cells. For modelling large scale microlattice structures, homogenisation is also an appropriate approach. This latter approach has been followed for foams for a number of years, but modelling localised plasticity, buckling and rupture is problematic. Analytic modelling of microlattice behaviour is useful for parametric investigation and to define and investigate specific structural cases. The synthesis of optimal microlattice structures is problematic given non linearities in response issues. Formal optimisation approaches are not currently possible, but the distinct approach of generative design is relevant. These approaches are discussed in terms of the specific structural applications of interest in this book.

Keywords Finite element modelling · Analytic modelling · Lattice synthesis · Lattice optimisation · Voxel modelling

5.1 Finite Element Modelling—Beam Elements

Gümrük and Mines (2013) developed a strut beam finite element analysis model for stainless steel 316L BCC microlattice block crush. Figure 5.1 gives details of their model. The finite element analysis program used was LS-DYNA, as the ultimate aim was to model block impact. A feature of the use of beam elements (Hughes-Liu) for microlattices is inaccurate modelling of stiffness at the nodes. This has been discussed by Luxner et al. (2005) and they proposed an increase of beam element stiffness in

© The Author(s), under exclusive licence to Springer Nature Switzerland AG 2019
R. Mines, *Metallic Microlattice Structures*, SpringerBriefs
in Structural Mechanics, https://doi.org/10.1007/978-3-030-15232-1_5

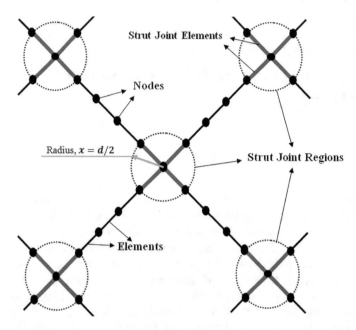

Fig. 5.1 Beam based strut finite element model for SS316L block. Cell size = 2.5 mm. Beam element properties adjacent to nodes are modified

the vicinity of the nodes. Gümrük and Mines (2013) increased the stiffness of the beam elements in the vicinity of the node by 50%. Another issue is the number of beam elements for a strut, and Gümrük and Mines (2013) proposed an element aspect ratio of 1, i.e. six elements for a 1.25 mm long strut. The model was applied to block crush (stainless steel 316L, BCC) and deformation details similar to the experimental case occurred. It should be noted that the selection of the strut diameter and parent material properties were dependent on the microstrut, the microlattice manufacture and quality, as discussed in Chaps. 3 and 4.

Smith et al. (2013) addressed the beam stiffness and node issues by increasing the diameter of the strut adjacent to a node by 50%. They used ABAQUS standard software and two node beam elements. Ushijima et al. (2010) used struts with rotation constrained at the node, and similar software discretisation was used. All these analyses gave similar results, which compared to experiment. It should be noted that for block loading, boundary conditions were simple compression, and that, as the stainless steel 316L BCC is ductile, the blocks are stable in collapse. Luxner et al. (2009) applied this beam model approach to disordered microlattice struts.

Labeas and Sunaric (2008) conducted a detailed investigation of the strut finite element analysis beam element modelling, using ANSYS and three dimensional beam elements. They used one element per strut, and calculated stiffness, via E, from block experimental data. As far as nonlinear behaviour is concerned, they modelled

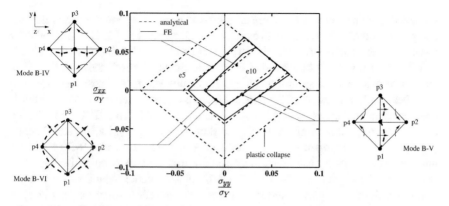

Fig. 5.2 Failure envelope for octet truss, showing plastic collapse (dashed lines) and buckling (solid lines) failure modes. Elastic buckling collapse governs collapse for all stress states other than bi-axial tensile stress state (which is plastic yield). Also shown are buckling collapse modes (reprinted from—Deshpande et al. (2001) with permission from Elsevier)

strut bending plasticity and compression buckling for the BCCZ topology. Six beam elements per strut were used.

Ptochos and Labeas (2012) extended their beam finite element analyses to block shear loading and they included shear effects in the beam elements (Beam188 in ANSYS). The effect of inclusion of shear was dependent on beam aspect ratio, and differences appeared at aspect ratios greater than 0.1.

Mohr (2005) idealised the octet truss unit cell with one rod element for a strut, giving 36 elements per unit cell. The author subjected the unit cell to compression, tension and shear. For shear, the author identified shear induced contraction and shear induced expansion.

Experimentally measured multi axial loading of microlattice blocks was discussed in Chap. 4. Ushijima et al. (2013) modelled the multi axial behaviour of stainless steel 316L BCC microlattice blocks using 20 beam elements for each strut in MSC Marc. They differentiated elastic buckling and plastic yielding. They developed yield surfaces under bi-axial stress.

Deshpande et al. (2001) modelled their octet truss with 20–30 beam elements for struts, and also investigated elastic buckling and plastic collapse. Figure 5.2 gives plastic and buckling collapse for the octet truss in two dimensional stress space. Different buckling modes are identified, and it is shown that buckling is the dominant failure mode.

As far as compression of microlattices under impact loading is concerned, standard finite element analysis codes will model inertial effects. The strain rate dependence of selected materials was discussed in Chap. 4.

Gümrük et al. (2018) took their stainless steel 316 L BCC beam element block model and included material strain rate effects. They used LS-DYNA 3D, six beam elements for struts and Mat 24. In this, the Cowper Symonds parameters were used to scale the complete stress strain curve. It should be noted that the beam element modelled growth of strut plasticity in an approximate manner, however finite element analysis results are consistent with experiment. Ozdemir et al. (2017) idealised their re-entrant microlattice struts with beam elements and they included strain rate effects (for Ti 64) using the Johnson Cook strain rate model. In both these studies, it is concluded that material strain-rate effects dominate over micro inertia effects.

Labeas and Ptochos (2015) conducted a finite element analysis of the sandwich panel studied by Mines et al. (2013), i.e. stainless steel 316L BCC Microlattice core with carbon fibre reinforced polymer skins (see Chap. 7). Labeas and Ptochos (2015) idealised microstruts with Hughes-Liu beam elements (with LS-DYNA). The number of elements for each strut increased towards the impact point, i.e. 1–3. The strut material was idealised by a linear elastic plastic model (Mat 03). An elongation to break of 60% was included. The core skin interface was modelled using a contact algorithm. Damage in the skin was modelled using Mat 54 (Tsai Hill composite failure model). Results were compared with experiment (see Chap. 7) and a homogenised core model (see later).

5.2 Finite Element Modelling—Solid Elements

For more detail on the collapse of microlattice blocks, struts can be modelled using three dimensional continuum elements. In this, only a single or a small number of cells can be modelled. Smith et al. (2013) modelled stainless steel 316L BCC microlattices using ABAQUS Standard 3D brick elements. Standard plasticity material models were used. Figure 5.3 shows the distribution of von Mises stresses for the cases of 1.25, 1.5, 2 and 2.5 mm cell size. This model can discriminate maximum plasticity due to bending of struts near the nodes. From experimental tests (Mines et al. 2013) it has been shown that, for Ti 64 blocks, they fail at low strain, and so there is a need to include material failure in the finite element analysis models. Ullah et al. (2016) have studied Kagome microlattice struts under compression and shear. Kagome microlattices are efficient for shear loading. The authors used ABAQUS explicit with tetrahedral prism and brick elements. They used the ABAQUS standard plasticity model. The main point of interest was the modelling of material rupture. They used failure strain vs. triaxialty data, to address scale and mode of loading as discussed in Chap. 2. Numerical and experimental results are given in Fig. 5.4, and also shown are the failure modes for the four topologies.

Li (2015) developed a 3D finite element analysis model of stainless steel 316L microlattice blocks using tetrahedral elements. The author included elastic plastic constitutive data using the Johnson Cook model. Areas of maximum plastic strain and rupture behaviour were identified.

Tancogne-Dejean et al. (2016) analysed octet truss microlattices using ABAQUS, using first order solid elements (C3D8R) and the J2 plasticity model. They do not model material failure. The authors showed good agreement between finite element results and experiment, and they quantified the effects of material strain rate for impact loading.

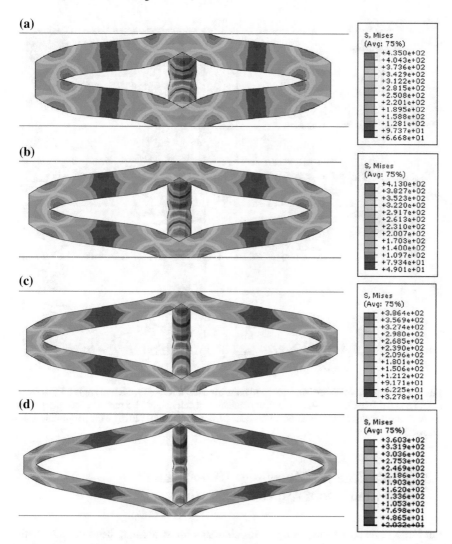

Fig. 5.3 Solid element finite element analysis of BCC SS316L microlattice block cell, for cell size **a** 1.25 mm, **b** 1.5 mm, **c** 2 mm, and **d** 2.5 mm. Contours shown for Von Mises stresses at 50% crush. Maximum stress is due to bending in micro struts adjacent to nodes (dark shaded area) (reproduced from Smith et al. (2013) with permission from Elsevier)

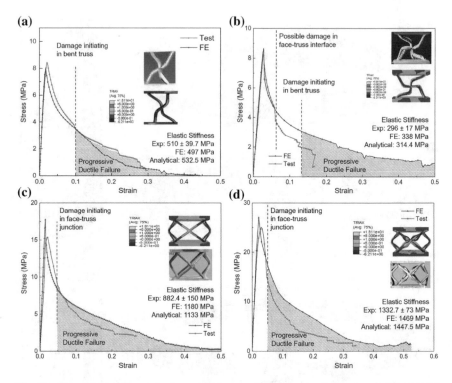

Fig. 5.4 Compressive failure behaviour and failure modes. **a** Kagome (internal angle 55°), **b** BCC, **c** F2CC, and **d** F2BCC with 0.81 mm strut diameter and cell height 11.5 mm. Contour plots show tri-axial stress state. Material: Ti 64 (reproduced from Ullah et al. (2016) with permission from Elsevier)

5.3 Finite Element Modelling—Homogenised and Continuum Approaches

For large scale structures, using beam elements for struts may become unrealistic and so continuum and homogenised approaches become appropriate. For stainless steel 316L BCC Microlattice structures used as core of sandwich panel subjected to foreign object impact (see Mines et al. 2013 and Shen et al. 2014), Labeas and Ptochos (2015) developed an homogenised approach. In this, for the elastic phase, they identified two independent elastic constants, for tension and shear. For nonlinear behaviour, they modelled compaction under compression or ultimate failure in shear. The homogenised model used Mat40 in Ansys (Mat_nonlinear.orthotropic) and LS DYNA. For the foreign object impact panel case, a quarter panel model was used, and numerical results were compared with experiment. Good agreement was obtained for impact force versus time behaviour. Also, deformation and damage in the model compared well with experiment.

Hundley et al. (2015) based their homogenised model for BCC aluminium 6101 on the work of Xue and Hutchinson (2004), for a square metallic honeycomb core. It is a simplified approach that neglects shear buckling. The orthotropic yield surface was modified to account for both the compression of the lattice and for the evolution of hardening. The theoretical model was implemented as vectorised user material subroutine (VUMAT) in ABAQUS. The model for foreign object impact of the sandwich panel was compared with experiment. They obtained good agreement with their force time plots. Good agreement also occurred for impact damage.

Mohr (2005) and Hu and Park (2013) developed homogenised models for the octet truss. Mohr (2005) considered first a single truss, and then extended the theory to a representative volume element, that comprised N truss members. The author addressed kinematics at finite strain, constitutive equations at microscale, and microscopic stress tensor for the small strain and finite strains. For the octet truss analysis, the small strain approach was taken. The behaviour of a unit cell was then discussed in detail. The approach was then applied to a two dimensional three point bend beam with five elements for depth and 20 elements for half span. Responses were compared with a discreet strut model. Hu and Park (2013) include multi axial effects in their model, and applied their model to an octet truss unit cell and uni-axial loading

5.4   Analytic Modelling of Microlattice Structures

Analytic modelling is useful for parametric variation, e.g. strut diameter, cell size or parent material properties.

Analytic modelling of the BCC stainless steel 316L microlattice structures have been developed for compression loading by Ushijima et al. (2010) and for multi axial loading by Ushijima et al. (2013). As far as compression is concerned, elastic behaviour was modelled using energy methods (Bending only) and the controlling equation derived for the stiffness of the block (E^*_{BCC}) was:

$$E^*_{BCC} = \pi\sqrt{3}E\frac{\left(\frac{d}{L}\right)^2}{1 + 2\left(\frac{L}{d}\right)^2}$$

where E is the modulus of the parent material, d is the diameter of the strut and L is the length of strut.

As far as plastic collapse is concerned, it was assumed that plastic collapse was controlled by plastic hinges in the vicinity of the nodes. Hence for collapse of the block, $\sigma^*_{pl,BCC}$:

$$\sigma^*_{pl,BCC} = \frac{4\sqrt{2}}{3}\sigma_s\left(\frac{d}{L}\right)^3$$

where σ_s is the yield stress of the parent material.

The range of limits of this approach was shown to be $0.05 < d/L < 0.17$, where d is the strut dimeter and L is the strut length. Hence, these parameters can be plotted with d/L. No optimum was shown, but at the extremities of d/L values, other failure modes will occur. Material strain rate effects could be included using the models discussed in Chap. 2.

These analyses were extended by Ushijima et al. (2013) to the multi axial loading case. The same assumptions and approach were made, namely bending only, energy approach for elastic response and plasticity near the nodes. They considered bi axial and tri axial stress states. They also included elastic buckling. The final results were expressed in terms of block stiffness and strength versus relative density, for the different angles in the BCC topology.

Deshpande et al. (2001) developed analytic failure surfaces for the octet truss. They used pin loaded joints for elastic response. For collapse, they investigated plasticity and buckling, and they included the effects of imperfections.

Doyoyo and Hu (2006) and Fan et al. (2008) developed failure envelopes for a variety of lattice topologies. Fan et al. (2008) included buckling failure modes, and derived specific uniaxial and shear strength values for four microlattices. Doyoyo and Hu (2006) identified strengthening and slenderness ratios as key design parameters.

5.5   Synthesis of Microlattice Topologies

Discussion so far has centred on analysis methods (numeric and analytic) for quantifying the dynamic nonlinear behaviour of volumes of microlattice structures. These methods take an existing states of affairs, and analyses them in order to gain further understanding of behaviours. However, Additive Manufacturing Technology allows realisation of more complex microlattice configurations. Hence, the techniques of topology optimisation are appropriate. However, given the complexity of the problem, e.g. multi axial, nonlinear, transient, multi component (core—skin), there is no complete generalisation at this time. Hence the first part of the discussion here focusses on specific cases. The second part of the discussion focusses on more general methods.

5.5.1   Optimisation for a Specific Stress State, e.g. Compression and Shear

Asadpoure and Valdevit (2015) developed an optimisation approach to minimise the weight of a two dimensional periodic lattice structure under simultaneous axial and shear stiffness constraints. As the optimisation process progresses, inefficient elements were eliminated and the cross sections of the remaining elements were resized. Compared to classic optimisation, the approach increased the computational cost by two orders of magnitude for the 2 dimensional case. The authors applied their approach to struts that are circular and hollow, and to locally and globally connected meshes. The authors compared their results with Hashin Shtrickman bounds and classic isotropic lattices e.g. hexagonal, fully triangular and Kagome.

A feature of the optimisation process is the ability to systematically tailor axial and shear modulus. Messner (2016) generated the optimisation approach to incorporate the full spectrum of lattice topologies. The author also used a homogenised material model with a parameterised description of the design space, and compared results with the Hashin Shtrickman bounds. It is proposed that the iso strut structure outperforms the octet truss, but requires a realisation process that can vary the size of the strut crossection.

5.5.2 Optimisation for a Specific Structural Element, e.g. Three Point Bend Beam

Stankovich et al. (2015) applied the optimality criteria process to the periodic lattice cells of specific geometry in cantilever beams. The optimality criteria method was preferred to stochastic methods and least squared fitting methods. They optimised cell size for two different materials and strut sizes for minimum weight for a given beam deflection.

Harl et al. (2017) used configurational based optimal design for a beam under 3 point bend. They looked at four different configurations and related optimal topologies to conventional lattice structures. The authors compared their theoretical results to experimental results for sintered polyamide, for linear beam displacement and concluded that for this specific case, the fully optimised framework is superior to periodic lattices.

5.5.3 Optimisation for a Combination of Structures, e.g. Core Skin

Tang and Zhao (2014) applied optimisation methods to lattice skin structures. Instead of optimising the topology of each lattice unit, the topology optimisation method is used to optimise material distribution and the whole design domain for two dimensional lattice structures. A hybrid approach was used, in which an initial design was taken, optimised and overall materials distribution optimised. The authors applied the methods to a sandwich beam under four point bend.

5.5.4 Optimisation for Specific Applications, e.g. Distributed Load in Bracket

So far in the discussion, simple lattice material volumes and boundary conditions have been considered as these are tractable for rigorous theoretical analysis. However, from an application point of view, there is a need to synthesise more complex configurations and actual components. For example, Chang and Rosen (2013) developed a methodology for optimising a three dimensional L bracket beam (see Fig. 5.5). They proposed a size matching and scale (SMS) method. This is similar to performance driven generative design methods discussed by Shea et al. (2005). The design method is the follows:

(a) specification of initial conditions, (b) generation of a ground geometry and unit cell regions, (c) solid body geometric analysis, (d) stress correlation and normalisation, (e) unit cell topology generation, (f) ambiguity resolution, and (g) diameter sizing.

Step (g), diameter sizing, is the most time consuming (80–90% of the total SMS design time). The method was used to optimize the bracket in Fig. 5.5 with constraints of minimum deflection and a given volume. This is a linear problem. Also, the SMS approach cannot deal with conformal configurations. However Nguyen et al. (2013) applied the method to the design of a micro aerial vehicle.

5.5.5 Optimisation for Non Linear Behaviour

Osanov and Guest (2016) discussed topology optimisation for architected materials design. They highlighted topology optimisation for material architectures governed by nonlinear mechanics. They discussed issues associated with homogenisation for a large number of cells, prior to optimisation. They referred to a paper by Carstensen et al. (2015). Carstensen et al. (2015) developed an approach which split elastic properties (unit cell) from nonlinear properties (structure with finite period density). They applied the method to the compression of a lattice block made from bulk metallic glass and for a lattice volume of 5×5 cells. The bilinear stress-strain for the parent material was assumed. They optimised the unit cell for a 12.5% volume fraction case. They compared their structure to equivalent cellular honeycombs, and showed that the energy absorption is 10 times higher. Alberdi and Khandelwai (2018) used computational homogenisation and Hoffman plasticity to investigate the optimisation of 2D lattice blocks under compression and shear. The authors identified a number of optimised topologies, and highlighted that future work is required on multi materials, more complex deformation modes and multi scale design.

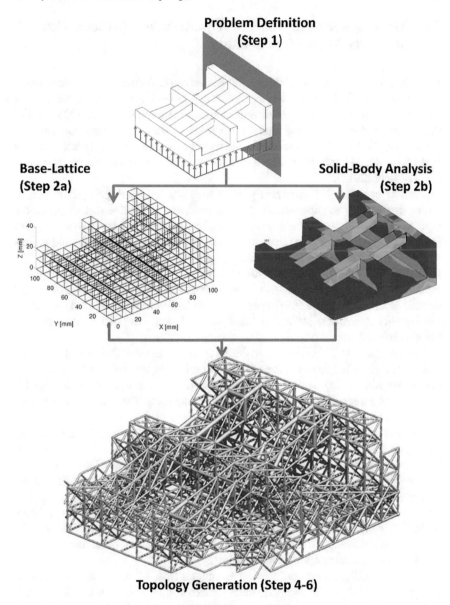

Fig. 5.5 Optimised lattice mesh for distributed load on a bracket (reprinted from Chang and Rosen (2013) with permission from Taylor and Francis)

5.6   More General Approaches: Optimisation Methods, Use of Voxels, Multi-Functionality

From the previous discussion, lattice optimisation can be focused on specific (simple) loading or on heuristic approaches for more complex loading case. A major theme for the latter case is the generation of complex lattice structures (Shea et al. 2005). In this, the structural shape annealing method was used, in which near randomly selected rules are applied in near randomly selected locations. In the method, the user controls rule selection and application. The technique is widely used in architectural design, e.g. Tedeschi (2014).

Panesar et al. (2018) discussed general methods for topology optimisation of lattice structures. The paper focussed on the realisation of lattice structures that are derived from topology optimisation solutions. The authors considered a wide range of lattice topologies, namely (a) truss based, and (b) surface based—solid and hollow shells. The surface based lattice structures are discussed further in Chap. 7. The authors considered the linear response of a cantilever beam. The authors integrated manufacturing processes into their design approach. The authors concluded that optimized lattices are 40–50% superior to standard uniform lattices. The authors gave quality descriptors indicating 'most to least' promising design strategies for several objectives.

Tamburrino et al. (2018) reviewed design methods for meso (micro) lattice structures. They identified design phases as unit cell selection, unit cell sizing, population of design space, setting density gradient and verification and optimisation. They differentiated between advanced and emerging technologies. Their review was mostly focused on topology and lightweight structures.

Aremu et al. (2017) discussed voxel based methods for representing lattices structures, and they included the feature of net skins. In this technique, geometry was represented by small blocks in three dimensional space. The size of the blocks need to represent all the structural details. The conversion of block cells to additive manufacture files is dependent on the additive manufacturing process used e.g. selective laser melting or binder jetting (Aremu et al. 2017). The approach has the advantages of flexibility of use, but disadvantages of resolution and rigour in optimisation techniques. The technique still needs to be further developed, and to be integrated into additive manufacturing realisation processes.

Chen et al. (2018) used optimisation techniques (method of moving asymptotes) and voxel based representation to generate extremal microstructure families. In this technique, single cells were discretised by $64 \times 64 \times 64$ cells. The authors used a sampling algorithm to reduce the number of combinations, and they used nonlinear dimensional embedding space and Isomat methods to investigate topology and associated properties. They focussed on elastic behaviour, and they validated their results with blocks made using an EOS SLS printer and PEBA 2301 material, and subjected to compression loading. They compared BCC microlattices with more complex microlattices, e.g. auxetic and chiral. As expected, the simple BCC maps on to the lower performance segment of the design space.

It should be noted that the BCC, BCCZ and OT geometries have been the main focus of attention in this book, given that there is extensive literature associated with these topologies, and that their behaviour is predictable for transient nonlinear behaviour. These topologies as can be used as benchmarks for the optimisation methods discussed.

Generalised optimisation can also tackle multi functionality. Ro and Roper (2017) discussed the surface area for photopolymer wave guides and electroless plated lattices, and highlighted the geometric relation between structural performance, and fluid flow and heat transfer. The authors developed analytic equations for solid and hollow lattices, volume fractions, and compactness equations for square, rectangular and hexagonal prisms. Volume fraction is important for mechanical strength and acoustics. Compactness is important for diffusion based processes, such as heat and mass transfer.

Valdevit et al. (2011) gave a wide ranging review of lattice cellular structures as multi functional devices. The authors discussed structural strength, heat transfer, thermo mechanical behaviour and fluid flow. Xiong et al. (2015) discussed multi functional aspects in their review of microlattice structures. They covered thermal applications, electrochemical applications, energy absorption, liquids absorption and biomedical applications. Jafari and Wits (2018) gave an overview of possible heat transfer devices using selective laser melting, and their paper showed the wide variety of solutions. Hence, it can be concluded that analyses need to be developed to formally relate functions, so that optimal multifunctional solutions can be formally developed.

5.7 Lattice Generation Software

Given the potential for realizing complex lattice structures using additive manufacture, there is significant activity in developing lattice generation software packages. Table 5.1 compares some of these, with some basic functionalities: e.g. generation, optimisation and manufacture.

As far as generation is concerned, issues include using standard cell types versus full design freedom, and inclusion of surfaces. Few software packages allow optimisation internally, and the usual method is to export the lattice files to a finite element analysis. Most software generate STL files, but these have to be converted for specific additive manufacturing machines. Other functions to be considered include generation of support structures and being open source. Few software packages currently integrate all realisation activities.

Table 5.1 Selected lattice generation software (As of Jan 2019)

Name	Web address	Comments
nTopology	www.ntopology.com	Stochastic./conformal topology, export CAD, free and professional
Betatype	www.betaty.pe	Manufacturing optimisation, PBF processes
Meshify	www.adimant.com	Upload geometry, generate print file, topology optimisation
Altair optistruct	www.altairhyperworks.com	Part of a large structural analysis suite. Powerful optimisation
Within (autodesk)	www.autodesk.com	Sophisticated, aimed at medical applications
Netfabb (autodesk)	www.netfabb.com	Lattice + surface, design own cells, standard/premium/ultimate versions
Flatt pack	www.flattpack.com	Surface lattice, developed by University of Nottingham, part of FLAC project
Materialise (magics)	www.materialise.com	Large library of cell types, interface modelling, lightweight struct. Modelling
Crystallon	www.food4rhino.com	Different cell types, generate print file, open source, similar to intralattice
Intralattice	www.intralattice.com	Different cell types, interface with FEA, generate CAD files, similar to crystallon
K3D surf	www.k3dsurf.sourceforge.net	Generates surfaces, generates file formats
Solidworks (3DXpert)	www.3dsystems.com	Conformal lattices, Use FEA, generate STL files, 5 versions
Simpleware	www.synopsis.com	Models lattice-skin, Use FEA, generate STL files, deals with TPMS
Frustum	www.frustrum.com	Surface lattices, A PTC company, uses generative design
Paramatters	www.paramatters.com	Generative design, built in FEA
Matlab (mathworks)	www.uk.mathworks.com	Standard cell, export FEA, 3D beam mesh, a plug in for Matlab, STL files
Monolith (autodesk)	www.monolith.zone	Voxel based, integral, STL/WRL/3DM file formats, free, multi material

5.8 Conclusions

Numerical and analytic solutions are appropriate for gaining further understanding of microlattice behaviours. Given the complexity of lattice structures, simplified numerical models are required—especially when transient nonlinear behaviour is being considered. Generalised synthesis and optimisation methods are being tackled, and these include additive manufacturing manufacturability. No generalised and formal approach is currently possible given the wide variety of topologies, parent materials, additive manufacturing processes and structural response. However, such approaches can be used to investigate the design space. Given the many outstanding issues in theoretical modelling, Chap. 7 discusses the experimental nonlinear behaviour of a number of lattice topologies, by solving specific application cases. Such experimental data can then be used to validate theoretical approaches and to benchmark for more optimal solutions.

References

R. Alberdi, K. Khandelwal, Design of periodic elasto plastic energy dissipating microstructures. Struct. Mult. Opt. (2018). https://doi.org/10.1007/s00158-018-2076-2

A.O. Aremu, J.P.J. Brennan Craddock, A. Panesar et al., The voxel based method of constructing and skinning conformal and functionally graded lattice structures suitable for additive manufacturing. Addit. Manuf. **13**, 1–13 (2017)

A. Asadpoure, L. Valdevit, Topology optimisation of lightweight periodic lattices under simultaneous compressive and shear stiffness constraints. Int. J. Sol. Struct. **60–61**, 1–16 (2015)

J.V. Carstensen, R. Lotfi, J.K. Guest, et al., Topology optimisation of cellular materials with maximised energy absorption, in *Proceedings of ASME 2015 International Design Engineering Technical Conferences and Computers and Information in Engineering Conference IDETC/CIE*, Boston, Massachusetts, USA, pp. 1–10, 2–5 Aug 2015 (2015)

P.S. Chang, D.W. Rosen, The size matching and scaling method: a synthesis method for the design of mesoscale cellular structures. Int. J. Comp. Integ. Manuf. **26**(10), 907–927 (2013)

D. Chen, M. Skouras, B. Zhu, et al., Computational discovery of extremal microstructure families. Sci. Adv. **4**, eaao7005 (2018)

V.S. Deshpande, N.A. Fleck, M.F. Ashby, Effective properties of the octet truss lattice material. J. Mech. Phys. Sol. **49**, 1747–1769 (2001)

M. Doyoyo, J.W. Hu, Multi axial failure of metallic strut lattice materials composed of short and slender struts. Int. J. Sol. Struct. **43**, 6115–6139 (2006)

H.L. Fan, D.N. Fang, F.N. Jing, Yield surfaces and micro failure mechanism of block lattice truss materials. Mat. Des. **29**, 2038–2042 (2008)

R. Gümrük, R.A.W. Mines, S. Karadeniz, Determination of strain-rate sensitivity of microstruts manufactured using the selective laser melting method. J. Mat. Eng. Perf. (ASM) **27**(3), 1016–1032 (2018)

R. Gümrük, R.A.W. Mines, Compressive behaviour of stainless steel microlattice structures. Int. J. Mech. Sci. **68**,125–139 (2013)

B. Harl, J. Predan, N. Gubeljak et al., On configuration based optimal design of load carrying lightweight parts. Int. J. Simul. Model **16**(2), 219–228 (2017)

J.W. Hu, T. Park, Continuum models for the plastic deformation of octet truss lattice materials under multi axial loading. J. Eng. Mat. Tech. (ASME). **135**, 021004—1–11 (2013)

J.M. Hundley, E.C. Clough, A.J. Jacobsen, The low velocity impact response of sandwich panels with lattice core reinforcement. Int. J. Imp. Eng. **84**, 64–77 (2015)

D. Jafari, W.W. Wits, The utilization of selective laser melting technology on heat transfer devices for thermal energy conversion application: a review. Renew. Sust. Energy Rev. **91**, 420–442 (2018)

G. Labeas, E. Ptochos, Homogenization of selective laser melting cellular material for impact performance simulation. Int. J. Struct. Integrity **6**(4), 439–450 (2015)

G.N. Labeas, M.M. Sunaric, Investigation on the static response and failure process of metallic open lattice cellular structures. Strain **46**(2), 1–10 (2008)

P. Li, Constitutive and failure behaviour in selective laser melting method stainless steel microlattice structures. Mat. Sci. Eng. A **622**, 114–120 (2015)

M.H. Luxner, J. Stampfl, H.E. Pettermann, Finite element modelling concepts and linear analyses of 3D regular open cell structures. J. Mat. Sci. **40**, 5859–5866 (2005)

M.H. Luxner, A. Woesz, J. Stampfl et al., The finite element study on the effects of disorder in cellular structures. Acta Biomater. **5**, 381–390 (2009)

M.C. Messner, Optimal lattice structured materials. J. Mech. Phys. Sol. **96**, 162–183 (2016)

R.A.W. Mines, S. Tsopanos, Y. Shen et al., Drop weight impact behaviour of sandwich panels with metallic microlattice cores. Int. J. Imp. Eng. **60**, 120–132 (2013)

D. Mohr, Mechanism based a multi surface plasticity model for ideal stress lattice materials. Int. J. Sol. Struct. **42**, 3235–3260 (2005)

J. Nguyen, S. Park, D. Rosen, Heuristic optimisation method for cellular structure design of lightweight components. Int. J. Prec. Eng. Manuf. **14**(6), 1071–1078 (2013)

M. Osanov, J.K. Guest, Topology optimisation for architected materials design. Am. Rev. Mat. Res. **46**, 211–233 (2016)

Z. Ozdemir, A. Tyas, R. Goodall et al., Energy absorption in lattice structures in dynamics: nonlinear FE simulations. Int. J. Imp. Eng. **102**, 1–15 (2017)

A. Panesar, M. Abdi, D. Hickman et al., Strategies for functionally graded lattice structures derived using topology optimisation for additive manufacturing. Addit. Manuf. **19**, 81–94 (2018)

E. Ptochos, G. Labeas, Shear modulus determination of cuboid metallic open lattice cellular structures by analytical, numerical and homogenisation methods. Strain **48**, 415–429 (2012)

C.J. Ro, C.S. Roper, Analytical models of the geometric properties the solid and hollow architected lattice cellular materials. J. Mat. Res. **33**(3), 264–273 (2017)

K. Shea, R. Aish, M. Gourtovaia, Towards integrated performance driven generative design tools. Autom. Constr. **14**, 253–264 (2005)

Y. Shen, W. Cantwell, R. Mines et al., Low velocity impact performance of lattice structure core based sandwich panels. J. Comp. Mat. **48**(25), 3153–3167 (2014)

M. Smith, Z. Guan, W.J. Cantwell, Finite element modelling of the compressive response of lattice structures manufactured using the selective laser melting technique. Int. J. Mech. Sci. **67**, 28–41 (2013)

T. Stankovic, J. Mueller, P. Egan, et al., A generalized optimality criteria method for optimisation of additively-manufactured multi material lattice structures. J. Mech. Des. (ASME) **137**, 111705-1–12 (2015)

F. Tamburrino, S. Graziosi, M. Bordegoni, The design process of additively manufactured meso scale lattice structures: a review. J. Comput. Info. Sci. Eng. **18**(4), 040801-1–16 (2018)

D. Tancogne-Dejean, A. Spierings, D. Mohr, Additively manufactured metallic microlattice materials for high specific energy absorption under static and dynamic loading. Acta Mater. **116**, 14–28 (2016)

Y. Tang, Y.F. Zhao, Design method for lattice skin structures fabricated by additive manufacturing, in *Proceedings of ASME 2014 International Mechanical Engineering Congress and Exposition IMECE*, Montreal, Quebec, Canada, pp. 1–9, 14–20 Nov 2014 (2014)

A. Tedeschi, *AAD—Algorithms aided design: parametric strategies using Grasshopper* (Le Penseur Publisher. Potenza, Italy, 2014)

I. Ullah, M. Brandt, S. Feih, Failure and energy absorption characteristics of advanced 3D truss core structures. Mat. Des. **92**, 937–948 (2016)

K. Ushijima, W.J. Cantwell, D.H. Chen, Prediction of the mechanical properties of microlattice structures subjected to multi axial loading. Int. J. Mech. Sci. **68**, 47–55 (2013)

K. Ushijima, W.J. Cantwell, R.A.W. Mines et al., An investigation into the compressive properties of stainless steel microlattice structures. J. Sand. Struct. Mater. **13**(3), 303–329 (2010)

L. Valdevit, A.J. Jacobsen, J.R. Greer et al., Protocols for the optimal design of multi functional cellular structures: from hypersonic to micro architected materials. J. Am. Ceram. Soc. **94**(S1), 1–20 (2011)

J. Xiong, R.A.W. Mines, R. Ghosh et al., Advanced microlattice materials. Adv. Eng Mater. **17**(9), 1253–1264 (2015)

Z. Xue, J.W. Hutchinson, Constitutive model for quasi static deformation of metallic sandwich cores. Int. J. Num. Meth. Eng. **61**, 2205–2238 (2004)

Chapter 6
Photopolymer Wave Guides, Mechanical Metamaterials and Woven Wire Realisation Methods for Metallic Microlattice Structures

Abstract This chapter discusses two alternative methods for manufacturing microlattice structures. In the first method, ultra violet light is shone into a liquid photopolymer, and the liquid solidifies in the volume that the light beam has irradiated. This means that complex lattice structures can be created out of liquid polymer. This lattice can either be used to create solid metallic microlattice structures, using investment casting techniques, or the lattice can be electroless plated with nickel phosphorus alloy. In the latter case, the polymer core is then removed, to produce ultra lightweight hollow microlattices. This chapter discusses mainly the manufacture, materials and progressive collapse of the ultra lightweight, hollow, microlattices. Such structures are an important class of the emerging field of mechanical metamaterials, and the latter are briefly introduced. The second method discussed is woven metal. In this, the metal wire (of the order of 1 mm in diameter) is shaped in three dimensions, and touching nodes are soldered or brazed. Relative densities of 6–43% can be obtained. Highly complex lattice patterns can be obtained. Both methods can be used to create shell lattice (Shellular) structures. Photopolymer wave guides are discussed first.

Keywords Photo polymer wave guides · Electroless plating · Investment casting · Mechanical metamaterials · Shellular · Woven wire

6.1 Photopolymer Wave Guides

Schaedler et al. (2011) discussed the manufacturing process (see Fig. 6.1). A thiolene liquid photo monomer is exposed to columated ultra violet light (UV) through a patterned mask, producing an interconnected three dimensional photopolymer lattice. Typically, 1–4 mm lattice member length, and 100–500 μm strut diameters can be achieved. Microlattice topology is limited by the path of the UV light. The thickness of the lattices is currently limited by the dissipation of the UV light, and so typically is of the order of five cells (approximately 20 mm).

Jacobsen et al. (2008) gave more detail on the formation and quality of the solid polymer microlattices. They gave details of various microlattice topologies and the

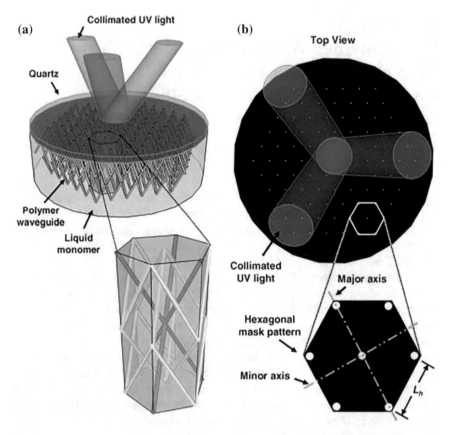

Fig. 6.1 Manufacture of microlattice using photopolymer wave guides. **a** Schematic of the setup for creating micro strut structures with an interconnected array of self propagating wave guides and **b** the top view of the mask with a hexagonal pattern of circular apertures (reprinted from Jacobsen et al. (2008) with permission from Elsevier)

failure of these under compressive loading. The authors gave details of the mask pattern that defines the topology of the polymer microlattices, and they realised two topologies using three and six exposure beams. They constructed 20 mm cubed blocks with 12 cells along the side. They compression tested these, and concluded that the six fold symmetry structures have 20–70% greater stiffness than the three fold symmetry structure.

The next step is to create hollow microlattices from the polymer lattice proformers. The electroless plating is achieved by immersing the polymer microlattice in a chemical bath, and the plating thickness is dependent on the time of immersion. The polymer core is then etched out, leaving hollow struts of wall thickness of 100–500 nm. This gives a microlattice density of 0.9 mg cm^{-3}.

As stated in Chap. 4, electroless plated nickel phosphorus can be an amorphous metal (metallic glass), hence such a material will behave differently from the more usual crystalline (structural) metal.

Sudagar et al. (2013) discussed electroless plating in detail and gave more detail of the acid bath nickel phosphorus alloy method. An important issue is the mechanical properties of the nickel phosphorus alloy, and Sudagar et al. (2013) quoted a yield stress of 900 MPa, a rupture strain of 0.7%, and a modulus of 100–120 GPa. Electro less plating can be achieved with widely available laboratory kits.

An important issue is the accuracy of the final hollow microlattices, given the ultra thin wall thickness. Salari-Sharif et al. (2018) gave a detailed analysis of this. Issues such as microstrut circularity, variation of wall thickness and node details need to be quantified. Variabilities, and imperfections, will affect the mechanical behaviour of the lattice blocks. Given the thin wall thickness, compressive block collapse will be dominated by elastic buckling. This should be contrasted with the thicker hollow struts discussed in Chap. 2, which typically fail due to plastic buckling.

Torrents et al. (2012) discussed the compressive behaviour and deformation modes of these microlattice structures in detail. They gave details of the nickel phosphorus alloy parent material. The blocks where $4 \times 4 \times 4$ cells, and a range of strut diameters, wall thicknesses and strut lengths were tested. All strut angles were at $60°$. Figure 6.2 gives a range of block compression stress-strain curves, showing loading and unloading characteristics. At ultralight densities, the blocks exhibit nearly full recovery at strains as large as 50%. At high densities (0.1–10%) the compressive behaviour is fully plastic, similar to traditional cellular materials. Figure 6.3 compares specific stiffness and strength for different relative densities. These values should be compared to selective laser melted Ti 64 BCC microlattice structures with a cell size of 2.5 mm: $E/E_s = 5.5 \times 10^{-4}$, $\sigma/\sigma_s = 8.3 \times 10^{-3}$ and $\rho/\rho_o = 0.060$. It is interesting to note that the step change in mechanical core mechanical performance is of interest to NASA in their development of Mars mission vehicles (NASA 2014). They are looking for a game change in mechanical properties of the order of 50% density reduction over state of the art.

Liu et al. (2014a) undertake a detailed analysis of the compressive behaviour of a microlattice configuration for energy absorption. The microlattices were realised using PPWG and electroless plating. They identified strategies for energy absorption enhancement, such as thickness gradients, radius gradients, and microlattice with water filler. The strategy is to enhance collapse modes, and hence overall energy absorption.

Liu et al. (2014b) extended their discussion to impact loading. They identified three regimes, namely, strain rate hardening (less than 50 ms^{-1}), inertial stabilisation (50–150 ms^{-1}) and shockwave (greater than 150 ms^{-1}). Hence, the dynamic behaviour of these microlattice structures can be adjusted by topology (cell size, strut diameter, wall thickness, strut diameter) and parent material.

It is interesting to note that photopolymer wave guides and electroless plating has been extended to integrated lattice core and skin sandwich panels (Choi et al. 2018). After manufacture, the polymer lattice core is sandwiched between glass slides. Further application of ultra violet light creates a solid polymer skin. The electroless

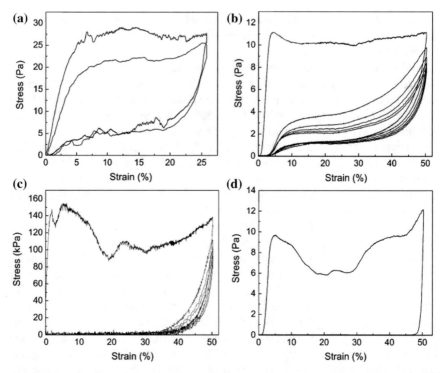

Fig. 6.2 Compression stress strain curves of hollow strut microlattice blocks manufactured using photopolymer wave guide. **a** t = 150 nm, strut diameter = 500 μm, l = 4 mm, density = 1.28 mg/cc. **b** t = 500 nm, strut diameter = 170 μm, l = 0.82 mm, density = 15.16 mg/cc. **c** t = 3 μm, strut diameter = 150 μm, l = 1.05 mm, density = 43.06 mg/cc. **d** t = 26 μm, strut diameter = 175 μm, l = 1.3 mm, density = 752 mg/cc (reprinted from Torrents et al. (2012) with permission from Elsevier) [Correction from original paper: for y axis: (a) × 10^{-3} P_a, (d) × 10^6 P_a]

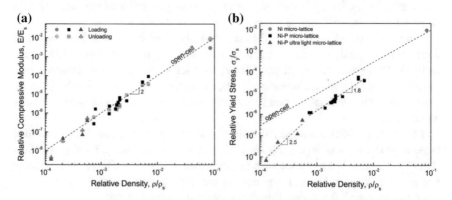

Fig. 6.3 **a** Relative compression modulus and **b** relative compression yield stress for microlattice blocks produced using photopolymer wave guide process for various wall thickness. Details given in Torrents et al. (2012) and reprinted with permission from Elsevier

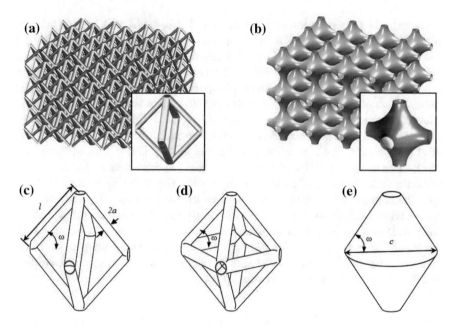

Fig. 6.4 Various shellular topologies. **a** BCC microlattice, **b** cubic shell, **c** detail BCC microlattice, **d** detail octahedral microlattice and **e** detail truncated conical shell (Reprinted from Lee et al. (2016) with permission from Elsevier)

plating is then applied as usual, and the polymer etched out. Resulting panels are nominally $14.5 \times 14.5 \times 4$ mm. Out-of-plane compression and in plane compression tests were conducted. The authors concluded that the skin needs to be at least four microns to be structurally effective.

The photopolymer wave guides and electroless plating method has been extended to shell like (Shellular) lattice structures (Lee et al. 2016; Han et al. 2015; Nguyen et al. 2016). This is achieved by changing the UV light mask, and Fig. 6.4 gives some geometries that can be achieved. A feature of standard microlattices are stress concentration at the nodes, and so the shell structures can dissipate these concentrations. Lee et al. (2016) discussed details of failure behaviour of these shell like structures using analytic and finite element solutions. Such structures are imperfection sensitive. The creation of micro and nano architected lattices materials using two photon polymerisation is discussed by Valdevit and Bauer (2016).

6.1.1 Multi Scale Metallic Microlattices

Zheng et al. (2016) extended the photo polymer wave guide manufacturing process to Large Area Projection Microstereolithography (LAPµSL), and used the process to create multi scale octet truss microlattices. In these, the Level 1 microlattice was approximately 2000 µm cell size with approximately 100 µm strut size. Each strut was then made from octet truss Level 2 microlattices, with cell size of the order of 100 µm and strut diameter of the order of 25 µm. A polymeric precursor was made, and then electro less plated with nickel phosphorus, giving hollow microlattices at Level 2. Cuboid blocks of $50 \times 20 \times 10$ mm ($40 \times 16 \times 8$ cells) were constructed with a fabrication time of 2 h/cm^3. Blocks of $3 \times 1.5 \times 1.5$ mm were tested under compression, shear and tension. Results were compared with first order (no hierarchy) octet truss blocks. For tensile loading, the hierarchical lattice showed increasing block strain to failure with similar specific strength to those with no hierarchy. The authors also studied tetrakaidecahedron topologies. Hence, it can be concluded that specialised multi scale metallic microlattices can be constructed (with restricted materials) to further populate the 'design/property space'. Mao et al. (2017) put the large scale area projection stereo lithography process in the context of other high resolution 3D printing processes. However, these processes usually use polymeric materials, and metallic lattices can currently only be achieved using the electroless plating process or the investment casting process.

6.1.2 Mechanical Metamaterials

A significant emerging field in structural mechanics is that of 'Mechanical Metamaterials'. Generally, metamaterials are materials that are artificially made, and which have properties not found in nature. This class of materials has already been touched upon in relation to ultra light weight microlattices and Shellular lattices, as discussed previously. In general, mechanical metamaterials use a number of emerging manufacturing processes, use a wide variety of materials (including non metallics) and typically are at nano scale. Given the fact that this book is focused on metallic additive manufacturing technology for conventional microlattice lightweight structures and functions, detailed discussion of mechanical metamaterials has not been given here.

However, Yu et al. (2018) provides a good starting point in the field. The authors discuss macro, micro and nano scale structures, a variety of mechanical properties (compression stiffness, bulk and shear moduli, Poisson's ratio), a variety of materials and a variety of manufacturing processes (additive manufacturing, interlocking assemblies, melt electro spinning techniques). Nonlinear structural behaviour of mechanical metamaterials was not discussed in detail. A brief mention is made of failure modes covering elastic buckling and parent material failure for selectively laser melted stainless steel microlattices, e.g. Li (2015). The authors also highlight the effect of reduction in geometric scale to enhance material and hence structural properties. However, the paper does introduce a large number of possible cellular topologies, and does address emerging issues, in Architectured Cellular Materials, of multiple materials and of multiple functions.

6.2   Woven Metal Wire

The other method to create solid microlattice structures uses metal weaving. Kang (2015) gave an extensive overview of this method. Conventional extruded wire (diameter of the order of 1 mm) was formed into various complex three dimensional shapes, and touching nodes are usually soldered or brazed. The technique requires complex forming and bending machines, and hence is a specialist activity. However, complex and large scale lattice structures can be built, unlike the other realisation methods discussed in this book. Kang (2015) gave details of the construction and testing of sandwich beams, brake disc core, multi functional panels and reinforced concrete structures.

Han et al. (2017) used the woven wire route to construct Shellular structures. In this, the wire is a polymer and the woven topologies are infiltrated with resin. The resin is subsequently cured, the resulting shell structure is then electroless plated and the polymer core etched out giving lightweight Shellular structures (wall thickness 1–$10\,\mu\text{m}$). The authors constructed $28 \times 28 \times 14$ mm blocks ($8 \times 8 \times 3$ cells), and compression test them. They showed that the one micron wall thickness blocks fail due to elastic buckling whereas the $10\,\mu\text{m}$ wall thickness blocks fail due to plastic buckling. The triply periodic minimal surface (TPMS) has mean curvature over the entire surface and so avoids the stress concentration found in strut based lattices. Also, such surfaces are appropriate for multi functionality, e.g. membrane for fuel cells or scaffolds in tissue engineering.

The realisation and properties of shell based lattice structures realised using selective laser melting and electron beam melting will be discussed in Chap. 7.

6.3   Conclusions

Thus to conclude, electroless plated microlattices manufactured using photopolymer wave guides extends possible microlattice architectures to ultra thin walled hollow struts microlattices and to shellular structures. The parent material is restricted due to the electroless plating process. New (complex) collapse behaviour mechanisms occur and behaviour is imperfection sensitive.

Woven metal wire methods require complex manufacturing machines, but allows realisation of industrially relevant configurations.

References

D.H. Choi, Y.C. Jeong, K. Kang, A monolithic sandwich panel with microlattice core. Acta Mater. **144**, 822–834 (2018)

S.C. Han, J.M. Choi, G. Liu et al., A microscopic shell structure with Schwarz's D surface. Nat. Sci. Rep. **7**, 13405-1–8 (2017)

S.C. Han, J.W. Lee, K. Kang, A new type of low density material: Shellular. Adv. Mater. **27**, 5506–5511 (2015)

A.J. Jacobsen, W.B. Carter, S. Nutt, Micro scale truss structures with three fold and six fold symmetry formed from self propagating polymer wave guides. Acta Mater. **56**, 2540–2548 (2008)

K.J. Kang, Wire woven cellular materials: the present and the future. Prog. Mater. Sci. **69**, 213–307 (2015)

M.G. Lee, J.W. Lee, S.C. Han et al., Mechanical analysis of "Shellular" an ultra low density material. Acta Mater. **103**, 595–607 (2016)

P. Li, Constitutive and failure behaviour in selective laser melted stainless steel for microlattice structures. Mater. Sci. Eng. A **622**, 114–120 (2015)

Y. Liu, T.A. Schaedler, A.J. Jacobsen et al., Quasi-static energy absorption of hollow microlattice structures. Compos. Part B **67**, 39–49 (2014a)

Y. Liu, T.A. Schaedler, X. Chen, Dynamic energy absorption characteristics of hollow microlattice structures. Mech. Mater. **77**, 1–13 (2014b)

M. Mao, J. He, X. Li et al., The emerging frontiers and applications of higher resolution 3D printing. Micromachines (MDPI) **8**, 113 (2017)

NASA, Game changing development programme, ultralight weight core materials for efficient load bearing composite sandwich structures. NASA Research Announcement Appendix No. NNH15ZOA 001N—15GCD—C1: Amendment 1 (2014)

B.D. Nguyen, J.S. Cho, K. Kang, Optimal design of Shellular: a micro architectured material with ultra low density. Mater. Des. **95**, 490–500 (2016)

L. Salari-Sharif, S.W. Godfrey, M. Tootkaboni et al., The effect of manufacturing defects on compressive strength of ultra light hollow microlattices: a data driven study. Addit. Manuf. **19**, 51–61 (2018)

T.A. Schaedler, A.J. Jacobsen, A. Torrents et al., Ultralight metallic microlattices. Science **334**, 962–965 (2011)

J. Sudagar, J. Lian, W. Sha, Electroless nickel, alloy, composite and nano coatings—a critical review. J. Alloys Compd. **571**, 183–204 (2013)

A. Torrents, T.A. Schaedler, A.J. Jacobsen et al., Characterisation of nickel based microlattice materials with structural hierarchy from nano metre to millimeter scale. Acta Mater. **60**, 3511–3523 (2012)

L. Valdevit, J. Bauer, Fabrication of 3D micro architected/nano architectured materials, in *Three Dimensional Micro Fabrication Using Two Photon Polymerisation: Fundamentals, Technology and Applications*, ed. by T. Baldacchini (Elsevier, Oxford, UK, 2016)

X. Yu, J. Zhou, H. Liang et al., Mechanical metamaterials associated with stiffness, rigidity and compressibility: a brief review. Prog. Mater. Sci. **94**, 114–173 (2018)

Z. Zheng, W. Smith, J. Jackson et al., Multiscale metallic metamaterials. Nat. Mater. **15**, 1100–1107 (2016)

Chapter 7
Applications for Additively Manufactured Metallic Microlattice Structures: Core Materials in Beams and Panels, Energy Absorbers (Static and Impact)

Abstract This chapter focuses on a small number of specific structural applications relevant to the additive manufacturing processes discussed. The focus here is mainly on two applications, namely core material in sandwich construction and energy absorbing devices. As far as core materials are concerned, two specific cases will then be discussed, namely cores in beam bending and in panels. In the case of energy absorbing devices, the use of lattices will be discussed first and then the use of more complex, shell like, elements will be discussed. The ultimate aim of this discussion is to quantify improvements in structural performance by the effective use of additive manufacture technology.

Keywords Core materials · Sandwich construction · Structural impact · Energy absorption · Surface based lattices

7.1 Core Materials in Beams

Some underlying structural ideas relating to truss core in sandwich constructions were discussed in Chap. 2 (Dragoni 2013). It was shown that for a specific topology, geometric parameters can be varied to optimise beam shear stiffness. Additive manufacturing enables the systematic variation of microlattice topology, cell size and parent material. An important element in the bending behaviour of sandwich construction is the properties of the skin, with the associated issues of skin core failure strain. Sandwich beams have multiple failure modes under three point bending (Mines et al. 1994), depending on the properties of constituent materials and on the structural loading conditions.

A specific beam bending case with a microlattice core was discussed by Shen et al. (2010). The cell was SLM body centred cubic stainless steel 316L microlattice, where the cell size was 2.5 mm. The skin was made up of four ply woven carbon fibre reinforced epoxy (EP 121 from Gurit A.G.). The skins were attached to the lattice core by heating at 225 °C in a hydraulic press, and maintaining the temperature for 1 h. Each ply was nominally 0.25 mm thick, giving a skin thickness of 1 mm. The core skin bond was achieved by pressing the microlattice into the skin. The core skin

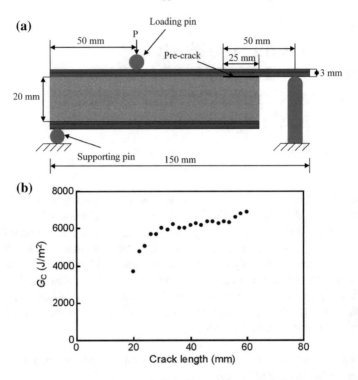

Fig. 7.1 Measured fracture data from SS316L BCC cored beam, with CFRP skins. **a** Beam test method, **b** fracture results (from Shen et al. 2014a, b). Reproduced with permission from Springer

bond strength has been studied in detail by Shen et al. (2014a, b). The fracture test specimen configuration and some results are shown in Fig. 7.1. A sandwich element was pre cracked along the bondline, and then the element was subjected to three point bending. Fracture toughness was derived from the force versus deflection data of the element, as the crack progressed in a stable manner. From these tests, it was concluded that the skin bond strength was high for both static and impact loading.

Sandwich beams were tested under three point bending, with spans of 80 mm (Shen et al. 2010, 2014a, b). Figure 7.2a shows force versus deflection data (loading rate of 4.2×10^{-6} ms^{-1}) and Fig. 7.2b shows the beam after failure. Beam initial failure can be a result of global bending or local deformation effects, which is dependent, in part, on the stiffness of the core. In this case, the beam progressively collapses with energy being absorbed by plastic deformation of microstruts (Shen et al. 2014a, b).

Shen (2009) compared SS316L BCC cored specific beam performance with titanium alloy Ti 64 microlattice cored beams and with metal foam cored (Alporas) beams, and the data is shown in Fig. 7.3. From this, it can be seen that the stainless steel 316L microlattice cored beam is comparable with aluminium foam case, and that the titanium alloy Ti 64 microlattice cored beam gives improved results. Hence, non

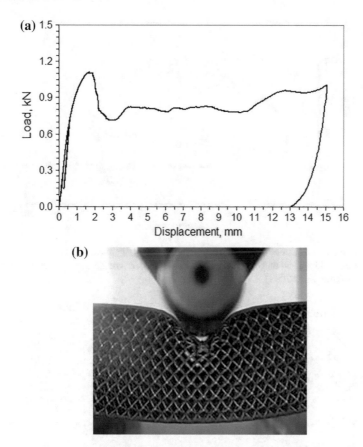

Fig. 7.2 Sandwich beam three point bend results, SS316L BCC core, CFRP skins. **a** Beam load displacement, **b** beam deformation after upper skin failure

optimized stainless steel 316L BCC Microlattices are comparable with aluminium foam (Alporas) as a core material in bending beams (with CFRP skin). Also, titanium alloy Ti 64 BCC microlattice core significantly increases beam stiffness and strength properties.

The full exploitation of additive manufacturing will enable systematic tailoring and optimisation of beam performance. In order to illustrate this, two examples with non metallic cores are discussed here.

Li and Wang (2017) investigated beams under three point bending, with three 2 dimensional core configurations (see Fig. 7.4). The cores were additively manufactured in an acrylic based photopolymer (Vero White) with a strut section of 1×1 mm thick. The core depth was 20 mm (four cell size) and the beam span was 60 mm. The skins were CFRP. Figure 7.5 shows beam failure modes. From the figure it can be seen that for the truss, overall core failure is due to shear. In the case of the conventional honeycomb, failure occurs across a layer in the beam. Finally, the re-entrant

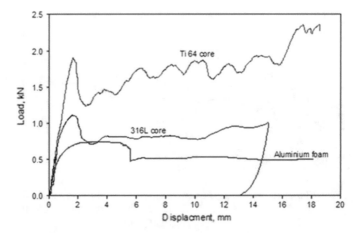

Fig. 7.3 Comparison of three point bend beam load deflection behaviour with three core materials and same CFRP skin. Ti 64 and SS316L lattices had BCC cell size 2.5 mm core, Aluminium foam was Alporas (from Shen 2009)

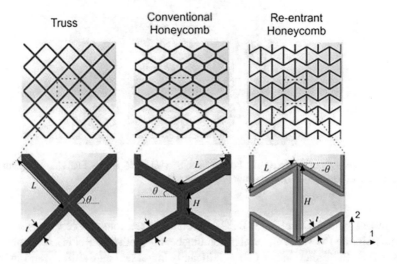

Fig. 7.4 Three core topologies from Li and Wang (2017). Material: acrylic based photopolymer (Verowhite), manufacture: material jetting. Cell size: 5.2–6.4 mm, strut thickness: 0.50–1.56 mm (reprinted with permission from Elsevier)

honeycomb also fails in a layered manner. Failure load reduction for the re-entrant honeycomb is less severe as compared to the other two cases. Finally, it was shown that for the intermediate relative density, the conventional honeycomb has the best performance, followed by the truss and then the re-entrant honeycomb.

An investigation of bending beams by Daynes et al. (2017), focussed on gradient lattice structures using Iso static lines. In this, non metallic ABS material was

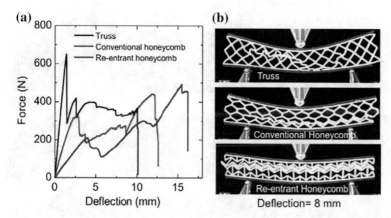

Fig. 7.5 Three point bend beam results for the three different core topologies. **a** force v deflection, **b** failure modes. Truss: failure load = 650 N, conventional honeycomb: failure load = 300 N, re-entrant honeycomb: failure load = 200 N (reprinted from Li and Wang (2017) with permission from Elsevier)

additively manufactured with a minimum strut diameter of 0.5 mm, and the cell length of 5 mm. The beam width was the same as the depth (30 mm) with a span of 150 mm. Figure 7.6 shows the optimised configuration. The authors changed both the diameter and spatial dimensions of the strut. There are seven layers through the thickness, linked by coarse cross bracing. There is no skin. The authors compared construction with standard periodic lattice structures. Experimentally, they showed a 30% increase in the stiffness for diameter graded and 100% increase in stiffness for spatially graded. It should be noted that manufacturability of graded configurations using metallic additive manufacturing could be problematic.

A major issue in the localised failure of sandwich beams is indentation failure at the loading point. Soden (1996) and Shueaib and Soden (1997) discussed this for conventional composite structures. Local beam failure can be modelled as:

$$P_f = \frac{4}{3}bt\sqrt{\sigma_s\sigma_c}$$

where P_f is the failure load, b is the beam width, t is the skin thickness, σ_s is the tensile strength of skin and σ_c is compression strength of the core (Soden 1996). Shuaeib and Soden (1997) compared theoretical predictions with experimental results for foam core and CFRP skin, and showed good agreement. Hence, it can be concluded that the core property, σ_c, controls beam failure behaviour and this can be adjusted with different microlattice structure topologies.

Thus to conclude this section on the specific case of beam bending, there are a number of strategies for optimising cell topology for improved beam stiffness, strength and energy absorption.

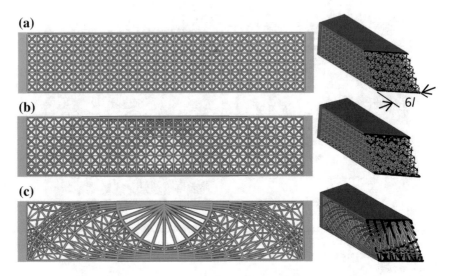

Fig. 7.6 Three different beam core configurations. **a** Non optimised, **b** strut diameter optimised, **c** spatial optimisation. Material: ABS, manufacture: material jetting, beam span: 150 mm, diameters strut 0.5 mm, length of struts 5 mm (reprinted from Daynes et al (2017) with permission from Elsevier)

7.2 Core Materials in Panels and Wing Sections

Discussion is now extended to sandwich panels. In this case, not only panel stiffness and strength are of importance but also foreign object impact becomes an issue, where panels are used in aerospace vehicles. The use of sandwich construction in commercial aviation was discussed by Herrmann et al. (2005). A major set of issues concerned foreign object impact from runway debris, bird strike, exploding tyres (Mines et al. 2007; Karagiozova and Mines 2007) and hail. The starting point for this work described in this book was foreign object impact behaviour of sandwich panels as discussed by Mines et al. (1998).

The stainless steel 316L selectively laser melted microlattice cell cored beam work (Shen et al. 2010) has been extended to panels (Mines et al. 2013 and Shen et al. 2014a, b). In these, the core and skin materials were the same. Figure 7.7 shows the experimental set up. The 100 × 100 mm panel was supported on spherical ball bearings in each corner, to simplify the computer simulation models. The impactor was of diameter of 10 mm, the drop weight mass was 2.07 kg, a maximum drop height was 1.5 m, giving a maximum impact energy of 23 J (Mines et al. 2013). Panels with stainless steel 316L BCC core were compared to panels with titanium alloy Ti 64 BCC cores, aluminium honeycomb cores and aluminium foam (Alporas) cores (Mines et al. 2013, Shen et al. 2014a, b).

Figure 7.8 shows damage for (a) SS 316L and (b) Ti 64 cores. There is a complex interaction between the skin and the core, and the lattice deforms and fails in a

Fig. 7.7 Microlattice panel
for drop weight loading.
BCC SS316L core (cell size
= 2.5 mm), CFRP skin.
Panel size 100 × 100 mm.
Distance between supports =
76 mm

complex manner. In order to gain further insights into panel behaviour, Mines et al. (2011) conducted a finite element analysis using LS DYNA. In this, the SS316L BCC core was modelled using three Hughes Liu beams per strut, bilinear elastic plastic model with failure (Mats 03) with properties: E = 140 GPa, σ_y = 250 MPa, E_T = 2.5 GPa, ε_f = 0.3. The latter was adjusted through comparison with experiment. Labeas and Ptochos (2015) analysed slightly larger panels, as discussed in Chap. 5. They used a similar finite element model, with one beam element per strut at the edges increasing to three beam elements per strut at the centre. Their material properties where E = 140 GPa, σ_y = 250 MPa, E_T = 450 GPa, ε_f = 0.16. The latter was assessed from micro strut tensile tests (Gümrük and Mines 2013). Such finite element analysis models the perforation of the panel and shows similar core failure modes to the experiment case. Validated finite element models can be used to virtually test for various parent materials, core topologies and panel boundary conditions.

Figure 7.9 compares perforation results. From these, it can be seen that the titanium alloy Ti 64 BCC and BCCZ cells are most comparable in perforation energy with aluminium honeycomb. It should be noted that in Mines et al. (2013) there were quality issues with the Ti 64 microlattice (Hasan et al. 2011), and it is believed that the Ti 64 performance could be improved (Mines et al. 2013). Shen et al. (2014a, b) studied, in detail, panel damage behaviour (skin and core).

Vrana et al. (2015) tested aluminium alloy AlSi12Mg BCC microlattice core panels under the drop weight impact loading, similar to Mines et al. (2013). Impact conditions where m = 5.75 kg, height = 1 m, diameter of impactor was 16 mm, panel area dimension was 45 × 45 × 16.8 mm deep. The skins where AlSi12Mg alloy and were integral to the panel. They varied strut diameters from 400 to 2000 μm. They studied damage behaviour in detail using CT scans. Generally, failure modes were similar to the Mines et al. (2013) and Shen et al. (2014a, b) work. Vrana et al. (2016) extended their studies to various microlattice topologies, viz. BCCZ, FBCC, FBCCZ

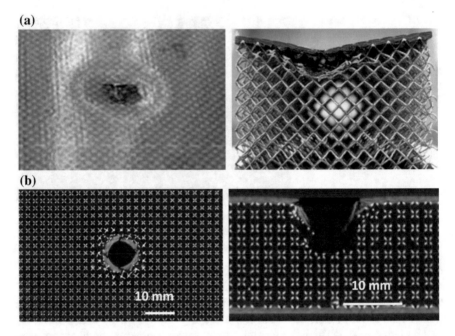

Fig. 7.8 Damage in sandwich panel shown in Fig. 7.7: **a** experiment (SS316L, impact energy = 8.8 J), **b** CT scan (Ti 64, impact energy = 23.0 J)

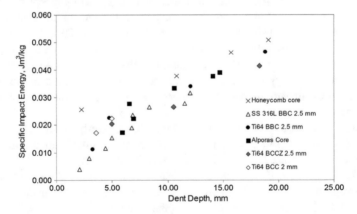

Fig. 7.9 Impact penetration results from panel tests from Mines et al. (2013) with panel geometry shown in Fig. 7.7

and gyroid. The authors showed that gyroid panel performance was similar to the BCC panel.

Abrate et al. (2018) studied BCC Ti 64 microlattices as core materials in sandwich panels with GRP skins. Panels were subject to drop weight loading. The drop weight was 6.5 kg, the impactor was 20 mm diameter, and impact velocities of 4, 5, 6 ms^{-1} giving an impact energy of 52, 81 and 117 J, respectively. The circular panel span was of diameter of 40 mm and was fully clamped. High energy impact gave rise to full skin perforation and core damage. Detailed failure mechanisms in the core were studied. Predictive impact models were applied to gain further understanding of impact mechanics. Extensive use of CT scans were used to give further data.

Drop weight impact panel tests have been conducted by Hundley et al. (2015), for microlattice cored panels, which were realized using the photo polymer wave guide approach (see Chap. 6). They considered BCC microlattices, manufactured using the investment casting approach. In this, the photopolymer was covered by ceramic slurry, the photopolymer was burnt out and the cavity filled with aluminium 6061. The surrounding coating was then removed. The diameter of struts were typically 200 μm. The panel core thickness was 12.7 mm, and panels size was 152.4 $\times$ 101.6 mm. The diameter of the impactor was 15.9 mm, and the impact mass was 14.4 kg. Impact velocity was up to 2.6 ms^{-1} (Height $= 0.35$ m), giving a maximum impact energy of 50 J. Skins where 1.45 mm thick 6951-O aluminium face sheets, which were brazed onto the core. Core damage was not discussed in detail, but the authors concluded that the lattice plasticity occurs, the core absorbs a portion of impact energy, and transfers loads between face sheets.

Yang et al. (2015) considered re-entrant lattice cores made from ABS and sintered (3D Systems Station Pro). Cell size was 4–7 mm, and strut diameter was 707–1000 μm. Block size was 50 $\times$ 50 $\times$ 30 mm, and the skins were of the same material manufactured integral with the core with thickness 1 mm. Impactor diameter was 19 mm, impact mass was 5.5 kg, maximum drop height was 0.025 m (velocity $= 0.7$ ms^{-1}) and maximum energy $= 1.3$ J. Panels with re-entrant (auxetic) microlattice cores were compared to panels with octahedral, rhomboid and hexagonal microlattices. From this work, it was concluded that the re-entrant (auxetic) microlattice materials gave higher average energy absorption and lower peak impact force, as compared with panels with other microlattice configurations.

Imbalzano et al. (2015) conducted similar work with re-entrant (auxetic) cores. This was purely a theoretical study on stainless steel 304 and aluminium 5083—H116 parent materials, panel sizes 300 $\times$ 300 $\times$ 30 mm. Cell size was 6.7 mm. Skins were the same materials as core and were 2 mm thick. An impactor of diameter 25 mm impacted with speeds of up to 150 ms^{-1}. Rate dependence of the parent material was modelled using the Johnson Cook model (see Chap. 2). The authors optimised re-entrant (auxetic) core microlattice parameters, and concluded that local deformation reduced by 56% for the auxetic cores versus equivalent solid (monolithic) plates.

In order to identify all the structural impact issues for low velocity impact of conventional sandwich structures, Chai and Zhu (2011) focussed on contact response, impact duration, large and small mass impact, impact velocity, impactor geometry, boundary conditions of panel and damage mechanisms. Core behaviour has significant effects on these issues, and additive manufacture cellular materials can be developed and optimized for specific behaviours.

To conclude this section. Additive manufacture for both polymeric and metallic materials, provide the opportunity to tailor and optimise panel (and especially foreign object impact) performance. Different topologies, parent materials, and cell scales are possible, along with different skin materials. Re-entrant microlattices give improved impact performance as compared with more conventional microlattices.

It should be noted that Smith (2012) extended discussion of stainless steel 316L BCC cores to the case of an idealised wing leading edge structure. The skin was the same as for panels (Mines et al. 2013). The leading edge core was created using the University of Liverpool Manipulator Software and was manufactured on a MCP Realiser II SLM 250 machine (Tsopanos et al. 2010). The CAD file of the core was cut by a surface representing the curved skin, hence no special steps were taken to optimise the core skin interface. The wing leading edge was manufactured using vacuum bagging in an oven. It was assumed that the core skin interface was similar to the panels (Mines et al. 2013). Drop weight impact conditions were the same as for the panel. Figure 7.10 shows quasi-static results with and without the core. From this, it can be seen that the effect of the core is to localise damage and reduce deformation. Of course, the stainless steel 316L microlattice adds mass, but this could be replaced by titanium alloy Ti 64 and could been reduced to a couple of cell thickness adjacent to the core.

Ullah et al. (2014) studied Kagome cores for aerospace wings. They focussed mainly on core stiffness and strength shear properties, and these are discussed in Chap. 4. They conducted detailed microlattice failure modelling. The kagome topology is advantageous for shear loading.

7.3   Energy Absorption in Solid and Hollow Strut Lattices

The static compressive behaviour of microlattices was discussed in Chaps. 4 and 6. Important issues highlighted included stable and progressive collapse modes, and good specific properties. Chapter 4 and Table 4.4 identified the effects on the lattice block crush of changing parent material, topology, and cell size.

Impact compression behaviour can be a result of low velocity impact (equipment/personal protection), foreign object impact (ballistic, debris, shrapnel) and blast loading. The impact dynamics of cellular materials has been studied extensively and some fundamental theories were discussed in Chap. 2. The two main phenomena are structural inertia and parent material strain rate effects.

Gümrük et al. (2018a) have studied the drop weight impact behaviour of stainless steel 316L BCC microlattices blocks, and they measured strain rate dependent material properties for single, and multiple strut tensile tests. They showed a 50% increase in block crush strain, which they attribute to material strain rate effects. They supported this conclusion by simulating the block tests using LS-DYNA with material strain rate been simulated using the Cowper Symonds (Perzyna) material

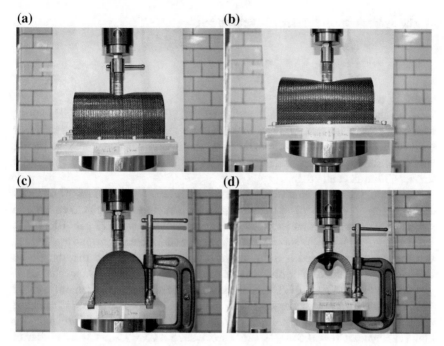

Fig. 7.10 Quasi static penetration tests for idealised wing leading edge. **a** Core, **b** no core, **c** core, and **d** no core (from Smith 2012)

model (see Chap. 2). It was also shown that block collapse failure modes did not change with impact loading.

McKown et al. (2008) and Smith et al. (2010, 2017) studied stainless steel 316L BCC and BCCZ microlattice structures under blast loading. In this, an explosive charge accelerated a mass of 0.2 kg up to 70 ms^{-1}, giving a maximum impact strain rate of 2000 s^{-1}. This mass crushed 20 mm cubed blocks. They expressed their results in terms of impulse (Ns), which can be related to impact velocity by changing momentum. They showed that the collapse mechanisms under blast was similar to the static case, and that an increase of 65–100% in crush stress occurred due to blast, which was mainly a result of material strain rate effects. A material strain rate of up to 1710 s^{-1} was found. McKown et al. (2008) also looked at the shock wave behaviour of these microlattices, and the critical velocity V$_{cR}$ is defined by:

$$V_{cR} = \left(\frac{2\sigma_{cr}^{qs} \varepsilon_D}{\rho} \right)^{0.5}$$

where σ_{cr}^{qs} is the quasi static collapse strength of the cellular material, ρ is the density of the cellular material, and ε_D is the densification strain.

The dynamic crush stress, σ_D is defined by:

$$\sigma_D = \sigma_{CR}^{qs} + \frac{\rho V_i^2}{\varepsilon_D}$$

where V_i is the impact velocity.

If the shock wave occurrences are defined as a 20% increase in the yield stress, then critical velocities are of the order of 30 ms^{-1}. However, shock wave evidence was not found from their experimental tests. The authors concluded that the impact was dominated by material strain rate effects, which is consistent with low speed impact (Gümrük et al. 2018a).

Lind et al. (2018) conducted high speed impact tests (up to 530 ms^{-1}) on small scale polymeric PMMA octet truss microlattice structures, manufactured using a stereolithography. Strut diameter was 21 μm and cell size was 177 μm. Block size was $1 \times 2 \times 3$ mm. The authors focussed on elastic lattice behaviour, and highlighted that the stress wave speed is independent of impact velocity, and that the elastic behaviour of the lattice conformed to classical theories of stress wave mechanics.

Ozdemir et al. (2016, 2017) studied the impact performance of titanium alloy Ti 64 microlattice block with a re-entrant topology (see Fig. 7.4). Impact was achieved using a metal Hopkinson bar set up bar ($V_i = 17$ ms^{-1}) and a Nylon 66 projectile for high velocity test ($V_i = 105$ ms^{-1}). Block size was $25 \times 25 \times 25$ mm, with cell size of 1.3 mm and a micro strut diameter of 480 μm. Specimens were manufactured using electron beam melting. It should be noted that the authors showed little material rate dependence for titanium alloy Ti 64 parent material.

Ozdemir et al. (2016) gave detailed collapse mechanisms and data for the blocks tested. The authors concluded that for Hopkinson bar tests, deformation was quasi static whereas for the Nylon 66 projectile test, shock wave effects occur, as discussed previously. Ozdemir et al. (2016) conducted a nonlinear finite element analysis of the tests. They conducted rate independent and rate dependent analyses. Their results confirmed the quasi static shock wave response as discussed above. They included material strain rate sensitivity in their models, and they concluded that this has little effect on block crush behaviours. The authors considered specimen confinement, and concluded that this has secondary effect on their re-entrant microlattices. However, the effect will be stronger for BCC microlattices as they have large lateral deformation during collapse (Gümrük and Mines 2013). McKown et al. (2008) proposed that lateral confinement for BCC will increase inertial effects due to buckling of microstruts.

Winter et al. (2014) studied the progressive collapse of selective laser melting microlattices at impact speeds up to 700 ms^{-1}. The microlattice geometry was BCCZ, with a cell size 1 mm and strut diameter of 400 μm. The material was stainless steel 316L. For these higher impact velocities, different physical phenomena appear, and so their simulation uses Eulerian Hydrocodes. They identified local deformation, which they called 'micro jetting', and this needs to be studied and optimised to improve impact performance.

Lijun and Weigdong (2018) systematically studied the impact dynamics of Ti 64 additively-manufactured selective laser melted rhombic duodecahedral microlattices, with graded density. The graded density was both stepwise and continuous. They used a split Hopkinson bar for the dynamic test. They identified static and dynamic block compression failure modes.

Tancogne-Dejean et al. (2016) discussed the impact loading of octet trusses. As discussed in Chap. 2, octet truss response is tension dominated and is more efficient compared to bending dominated structures e.g. BCC. Cell size was 3.1 mm with microstrut diameter of 500 μm and material stainless steel 316L manufactured using selective laser melting (Concept Laser M1). Block size was $7 \times 7 \times 7$ cells. Static tests were conducted at 1 mm per minute, and impact tests were conducted in the split Hopkinson bar ($V_i = 20$ ms^{-1}). They found a 30% increase in crush stress, and attributed this to parent material strain rate effects. Hence it can be concluded that for the octet truss, micro inertia effects are negligible for the two classes of impact considered.

Maskery et al. (2016) studied AlSi10Mg microlattices made using selective laser melting (Renishaw AM 250). Their cell sizes were 3 mm, their blocks $6 \times 6 \times 6$ cells, the microstrut diameter was 500 μm and the topology was BCC. The innovation here was graded blocks, along the line of compression. Cell size varied from for 4 mm at the top of the block to 2 mm at the bottom. Quasi static tests were conducted. They showed that the effect of grading is to increase energy absorption at larger deformations.

Liu et al. (2014) studied dynamic energy absorption characterisation of hollow microlattice structures manufactured using photopolymer wave guides. The microlattice can be idealised as BCCZ, with Z beam struts with larger diameter than the BCC element. Cell size was 10 mm, strut diameter varied from 2.4 mm (BCC) to 3.4 mm (Z), with wall thickness $= 40$ μm. The material was electro plated nickel. The block impact velocity was up to 150 ms^{-1}. They identified three regimes that were strain rate hardening (less than 50 ms^{-1}), inertial stabilisation (50–150 ms^{-1}) and shockwave (greater than 150 ms^{-1}).

Schaedler et al. (2014) compared a number of energy absorbing microlattice structures e.g. open cell metal foam, polystyrene foam, aluminium honeycomb, solid and hollow aluminium micro struts, and compared specific energy absorption and transmitted stress under compression. The authors concluded that microlattices offer more flexibility in tailoring responses to impulsive loads.

Gümrük et al. (2018b) take their standard stainless steel 316L BCC microlattices blocks, and electro plate them with Ni P. The thickness of the electroplating was 18 μm. They showed an increase of block specific stiffness by 75% and block specific strength by 50%. Block collapse modes were similar to the non electroplated case. Hence there is scope to optimise electroless plating thickness for optimised block properties. This is a specific example of multiple materials for porous structures and powder bed fusion processes.

To conclude this section, additive manufacturing microlattices offer many opportunities to tailor and optimise microlattice energy absorption performance. The selection of papers have been discussed here to highlight issues, and it can be seen that different manufacturing processes and materials give a wide variety of structural solutions. The next section extends discussion to configurations for energy absorption using surface based lattices.

7.4   Energy Absorption in Surface Based Lattices

The focus of this book has been on microlattice structures, as there is the largest amount of literature on this, and hence it is the most mature subject area. However, additive manufacturing can realise any three dimensional topology. Hence researchers are starting to study other micro topologies. Shell lattices where discussed in Chap. 6 relating to photopolymer wave guides and wire weaving. This chapter discusses shell lattices manufactured using additive manufacture.

For example, Bonatti and Mohr (2017) discussed stainless steel 316L selective laser melted micro shell topologies. They compared solid octet truss (OT), hollow sphere assembly (HSA), hollow octet truss (HOT) and hybrid truss assembly (HTA). Typically their cell dimensions were up to 10 mm, wall thickness were 320–350 μm, strut diameter (OT) was 1.3 mm, sphere diameter (HSA) was 5.15 mm, hollow octet truss (HOT) diameter was 2.17 mm and hybrid truss assembly (HTA) diameter was 5 mm. They focussed on hydrostatic compression, as this is important for cellular materials (Ashby et al. 2000). They attributed non idealised aspects of their realised structures to non wetted powder. They tested 5 × 5 × 5 unit cell blocks, typically 47.1 mm cubed. Large deformation responses were discussed in detail, along with block failure modes. They concluded that OT was the weakest and that HTA absorbed the most energy.

Maskery et al. (2017) studied AlSi10Mg double gyroid structures. The gyroid belongs to the family of triply periodic minimal surfaces (TPMS), which is a subset of a large class of constant mean curvature surfaces (CMCS). Blocks of 18 mm cubed were tested under compression, and failure modes identified. The authors gave the advantages of the gyroid topology as a flat crush stress plateau, and energy absorption was three times the case of the equivalent BCC structures.

Triply periodic minimal surfaces (TPMS) were first discussed be Schwarz (1856) and more recently by Schoen (2012). Minimal surfaces are defined as surfaces with zero mean curvature.

Al Ketan and Al Rub (2018) and Al Ketan et al. (2018) studied the mechanical properties of Triply periodic minimal surfaces (TPMS). Figures 7.11 and 7.12 give extracts from Al Ketan et al. (2018). In the latter paper, the authors considered standard lattices (Kelvin, octet truss, Gibson Ashby), skeletal TPMS (IWP, diamond, gyroid) and sheet DPMS (IWP, diamond, gyroid and primitive)—see Fig. 7.11. The authors conducted block compression tests in maraging steel manufactured by direct metal laser sintering (EOS INT M280) using contour and hatch laser scan strategies. The block size was 42 mm cubed, cell size was approximately 7 mm, wall thickness was typically 400 μm for shells and 800 μm for struts. The cell number was 6 × 6 × 6. The quality of the microlattices was good, with well defined and repeatable

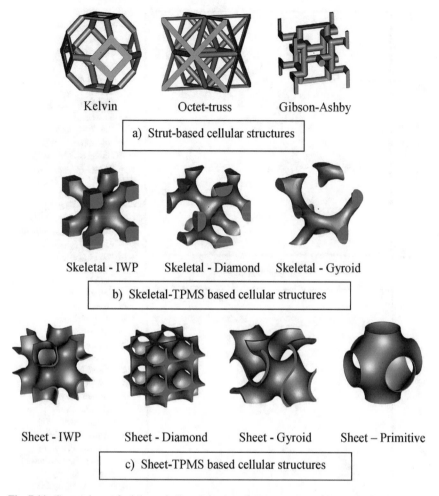

Kelvin   Octet-truss   Gibson-Ashby

a) Strut-based cellular structures

Skeletal - IWP   Skeletal - Diamond   Skeletal - Gyroid

b) Skeletal-TPMS based cellular structures

Sheet - IWP   Sheet - Diamond   Sheet - Gyroid   Sheet – Primitive

c) Sheet-TPMS based cellular structures

Fig. 7.11 Comparison of triply periodic minimal surface structures with standard microlattice topologies (reprinted from Al Ketan et al. (2018) with permission from Elsevier)

topologies. Figure 7.12 summarises some of the results. In general, the sheet diamond is the most efficient topology, with the skeletal IWP being one of the most inefficient. It should be noted that the standard BCC topology is typically 30% less stiff and 10% less strong as compared to the skeletal IWP.

Al Ketan and Al Rub (2018) includes BCC data. A feature of surface based lattice cell type structures is reduced stress concentration as compared with strut based structures, giving rise in general to smoother crush behaviours. The quality of these complex surface lattice structures will become an issue for lower density materials (e.g. Ti 64, AlSi10Mg) and for smaller features size. However, this approach gives another set of solutions for energy absorbing devices.

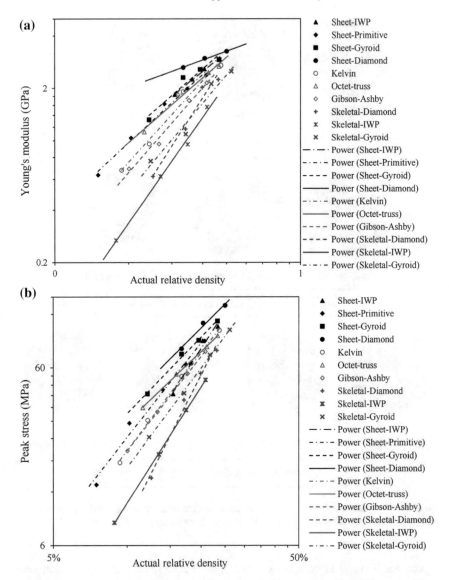

Fig. 7.12 Comparison of **a** modulus and **b** peak stress for compression of maraging steel lattice blocks with various topologies. BCC typically 30% less for Young's modulus, and 10% less for peak stress compared to skeletal IWP (reprinted from Al Ketan et al. (2018) with permission from Elsevier)

7.5 Quantification of Improvements in Structural Performance

We are now in a position to quantify the improvements in selected structural applications by using additive manufacturing technology, and associated strategies. Table 7.1 lists 10 examples for beams, panels and blocks (energy absorbers).

In case 1, the optimised tetrahedral topology improves beam shear strength over aluminium honeycomb. In case 2, Ti 64 SLM BCC core increases stiffness and strength over a beam with aluminium foam core. In case 3, the use of iso static lines in the core improves stiffness and strength over standard periodic cores. In case 4, for panel foreign object impact, the use of a Ti 64 core improves stiffness and strength over an aluminium honeycomb core. In case 5, the use of gyroid topology improves stiffness and strength over the BCC core. In case 6, the use of re-entrant (auxetic) topology improves strength over monolithic panel. In case 7, the use of Ti 64 instead of stainless steel 316L improves specific block energy absorption by 500%. In case 8, the use of a PPWG printed lattice improves energy absorption by 300% over aluminium foam. In case 9, the use of graded BCC improves energy absorption at large crush (approximately 40%). Finally, in case 10, the use of triply periodic minimal surfaces improves energy absorption by 400%. From these ten cases, it can be seen that structural behaviours are complex. These 10 cases give a number of solution strategies, independent of formal optimisation. Hence, the sophisticated mathematical methods discussed in Chap. 5 have some way to go in fully optimizing specific structural applications.

7.6 Conclusions

The focus for this chapter (and for this book) has been on three specific structural applications, namely sandwich beams, sandwich panels and energy absorbing devices. The aim here has been to allow in depth discussion of these three applications, given the complexity of the structural mechanisms, the transient and nonlinear behaviour, and of manufacture and materials issues. Lattice generation and optimisation approaches were discussed briefly towards the end of Chap. 5. Fairly obviously, additive manufacture has the potential to create highly complex lightweight structures and topologies. Hence, from a structural design point of view, there is huge potential to synthesise bespoke structures for a given application. This has not being addressed here. Specific performance improvements have been identified for beams, panels and energy absorbers for specific structural solutions. There is further scope to fully exploit additive manufacturing technology.

In order to widen discussion, the edited volume by Hamm (2015), gives a relevant overview of biologically inspired systems, that could be exploited using additive manufacture. At a more general level the classic text by D'Arcy Thompson (1961) (originally published in 1917) provides inspiration on biological topology.

Table 7.1 Summary of performance improvements of selected applications

Application	Reference	Material	Manufacture	Benchmark	Improvement	Stiffness	Strength	Density
Beam	Dragoni (2013)	Al Alloy	Conventional	Al honeycomb	Opt. tetrahedral	−24%[a]	+67%[b]	Similar
Beam	Shen (2009)	Ti 64 BCC/CFRP skin	SLM	Al Foam	BCC	+52%	+267%	+25%
Beam	Daynes et al. (2017)	ABS polymer	Material jet	Uniform BCC	Isostatic lines	DG: +30% SG: +100%	DG: +120% SG: +170%	Similar Similar
Panel	Mines et al. (2013)	Ti 64 BCC/CFRP skin	SLM	Al honeycomb	BCC	+290%	+93%	−400%
Panel	Vrana et al. (2016)	AlSi12Mg/skin + core	SLM	BCC	Gyroid	+17%	+17%	Similar
Panel	Imbalzano et al. (2015)	SS304, Al5083	Theoretical	Monolithic	Re-entrant/Auxetic	+56%	+100%	Similar
Block	Mines et al. (2013)	SS316L/Ti 64	SLM	SS316L	Ti 64	+50%	+500%	30%
Block	Schaedler et al. (2014)	Al Alloy	PPWG	Al foam	BCC	+300%	+300%	−100%
Block	Maskery et al. (2016)	AlSi10Mg	SLM	Periodic	Graded	Similar	+100%	Similar
Block	Al Ketan et al. (2018)	Maraging steel	SLM	BCC	Sheet diamond	+500%	+400%	Similar

Key [a]From shear modulus, [b]From shear strength, *DG* diameter graded, *SG* spatially graded

References

S. Abrate, G. Epasto, E. Kara et al., Computed tomography analysis of impact response of lightweight sandwich panels with microlattice core. Proc. Inst. Mech. Eng. Part C J. Mech. Eng. Sci. **232**(8), 1348–1362 (2018)

O. Al Ketan, R.K.A. Al Rub, The effect of architecture on the mechanical properties of cellular structures based on the IWP minimal surface. J. Mat. Res. **33**(2), 343–359 (2018)

O. Al Ketan, R. Rowshan, R.K.A. Al Rub, Topology mechanical property relationships of 3D printed strut, skeletal and sheet based periodic metallic cellular materials. Addit. Manuf. **19**, 167–183 (2018)

M.F. Ashby, A. Evans, N.A. Fleck et al., *Metal Foams: A Design Guide* (Butterworth Heinemann, Woburn, USA, 2000)

C. Bonatti, D. Mohr, Large deformation response of additively manufactured FCC metamaterials: from octet truss lattices towards continuous shell meso structures. Int. J. Plast. **92**, 122–147 (2017)

G.B. Chai, S. Zhu, A review of low velocity impact on sandwich structures. Proc. IMechE. Part L. J. Mat. Des. Appl. **225**, 207–230 (2011)

D'Arcy Thompson, *On Growth and Form* (Cambridge University Press, Cambridge, UK, 1961)

S. Daynes, S. Feih, W.F. Lu et al., Optimisation of functionally graded lattice structures using Iso static lines. Mater. Des. **127**, 215–223 (2017)

E. Dragoni, Optimal mechanical design of tetrahedral truss cores for sandwich constructions. J. Sand. Struct. Mater. **15**(4), 464–484 (2013)

R. Gümrük, R.A.W. Mines, S. Karadeniz, Determination of strain-rate sensitivity of microstruts manufactured using selective laser melting method. J. Mat. Eng. Perf. (ASM) **27**(3), 1016–1032 (2018a)

R. Gümrük, R.A.W. Mines, Compressive behaviour of stainless steel microlattice structures. Int. J. Mech. Sci. **68**, 125–139 (2013)

R. Gümrük, A. Usun, R. Mines, The enhancement of the mechanical performance of stainless steel microlattice structures using electroless plated nickel coatings. MDPI Proc. **2**(8), 494 (2018b)

C. Hamm (ed.), *Evolution of Lightweight Structures: Analyses and Technical Applications* (Springer, Dordrecht, 2015)

R. Hasan, R. Mines, P. Fox, Characterisation of selectively laser melted Ti 6Al 4V microlattice struts. Procedia Eng. **10**, 536–541 (2011)

A.S. Herrmann, P.C. Zahlen, I. Zuardy, Sandwich structures technology in commercial aviation: present application and future trends in *Sandwich Structures 7: Advancing with Sandwich Structures and Materials* by eds. O.T. Thomsen, et al., pp. 13–26 (2005)

J.M. Hundley, E.C. Clough, A.J. Jacobsen, The low velocity impact response of sandwich panels with lattice core reinforcement. Int. J. Imp. Eng. **84**, 64–77 (2015)

G. Imbalzano, P. Tran, T.D. Ngo et al., Three dimensional modelling of auxetic sandwich panels for localised impact resistance. J. Sand. Struct. Mater. **19**(3), 291–316 (2015)

D. Karagiozova, R.A.W. Mines, Impact of aircraft rubber fragments on aluminium alloy plates: 2 numerical simulation using LS DYNA. Int. J. Imp. Eng. **34**(4), 647–667 (2007)

G. Labeas, E. Ptochos, Homogenization of selective laser melted cellular material for impact performance simulation. Int. J. Struct. Integr. **6**(4), 439–450 (2015)

T. Li, L. Wang, Bending behaviour of sandwich composite structures with tunable 3D printed core materials. Comp. Struct. **175**, 46–57 (2017)

X. Lijun, S. Weidong, Additively-Manufactured functionally graded Ti6Al4V lattice structures with high strength under static and dynamic loading: experiments. Int. J. Imp. Eng. **111**, 255–272 (2018)

J. Lind, B.J. Jensen, M. Barham, et al., In situ dynamic compression wave behaviour in additively manufactured lattice materials. J. Mat. Res. (2018). https://doi.org/10.1557/jmr.2018.351

Y. Liu, T.A. Schaedler, X. Chen, Dynamic energy absorption characteristics of hollow microlattice structures. Mech. Mater. **77**, 1–13 (2014)

I. Maskery, N.T. Aboulkhair, A.O. Aremu et al., A mechanical property evaluation of graded density AlSi10Mg lattice structures manufactured by selective laser melting. Mater. Sci. Eng. A **670**, 264–274 (2016)

I. Maskery, N.T. Aboulkhair, A.O. Aremu et al., Compressive failure modes and energy absorption in additively-manufactured double gyroid lattices. Addit. Manuf. **16**, 24–29 (2017)

S. McKown, Y. Shen, W.K. Brooks et al., The quasi-static and blast loading response of lattice structures. Int. J. Imp. Eng. **35**, 795–810 (2008)

R.A.W. Mines, S. McKown, R.S. Birch, Impact of aircraft rubber tyre fragments on aluminium alloy plates: 1 experimental. Int. J. Imp. Eng. **34**(4), 627–646 (2007)

R.A.W. Mines, S. Tsopanos, S.T. McKown, Verification of a finite element simulation of the progressive collapse of microlattice structures. Appl. Mech. Mater. **70**, 111–116 (2011)

R.A.W. Mines, S. Tsopanos, Y. Shen et al., Drop weight impact behaviour of sandwich panels with metallic microlattice cores. Int. J. Imp. Eng. **60**, 120–132 (2013)

R.A.W. Mines, C.M. Worrall, A.G. Gibson, The static and impact behaviour of polymer composite sandwich beams. Composites **25**(2), 95–110 (1994)

R.A.W. Mines, C.M. Worrall, A.G. Gibson, Low velocity perforation behaviour of polymer composites sandwich panels. Int. J. Imp. Eng. **21**(10), 855–879 (1998)

Z. Ozdemir, E. Hernandez Nava, A. Tyas et al., Energy absorption in lattice structures in dynamics: experiments. Int. J. Imp. Eng. **89**, 49–61 (2016)

Z. Ozdemir, A. Tyas, R. Goodall et al., Energy absorption in lattice structures in dynamics: nonlinear FE simulations. Int. J. Imp. Eng. **102**, 1–15 (2017)

T.A. Schaedler, C.J. Ro, A.E. Sorensen et al., Designing metallic microlattices for energy absorbing applications. Adv. Eng. Mater. **16**(3), 276–283 (2014)

A.H. Schoen, Reflections concerning triply periodic minimal surfaces. Interface Focus **2**, 658–668 (2012)

F.M. Shuaeib, P.D. Soden, Indentation failure of composite sandwich beams. Comput. Sci. Technol. **57**, 1249–1259 (1997)

A.H. Schwarz, Gesammelte Mathematische Abhanglungen, vol. 1 (AMS Berlin, Germany, Springer, 1970) First published 1856

Y. Shen, W. Cantwell, E.Y. Li, Skin core adhesion in high performance sandwich structures. J. Zheijang Univ. Sci. A (Appl. Phys. Eng.) **15**(1), 61–67 (2014a)

Y. Shen, W. Cantwell, R.A.W. Mines et al., Low velocity impact performance of lattice structure core based sandwich panels. J. Compos. Mater. **48**(25), 3153–3167 (2014b)

Y. Shen, S. McKown, S. Tsopanos et al., The mechanical properties of sandwich structures based on metal lattice architectures. J. Sand. Struct. Mater. **12**, 159–180 (2010)

Y. Shen, High performance sandwich structures based on novel metal cores. Ph.D. Thesis, University of Liverpool, 2009

M. Smith, W.J. Cantwell, Z. Guan et al., The quasi-static and blast response of steel lattice structures. J. Sand. Struct. Mater. **13**(4), 479–501 (2010)

M. Smith, W.J. Cantwell, Z. Guan, Impact and blast response of lattice materials, in *Dynamics of Lattice Materials* by eds. A.S. Phani, M.I. Hussein. Wiley, Chichester, UK (2017)

M. Smith, The compressive response of novel lattice structures subjected to static and impact loading. Ph.D. Thesis, University of Liverpool, 2012

P.D. Soden, Indentation of composite sandwich beams. J. Strain Anal. **31**(5), 353–360 (1996)

D. Tancogne-Dejean, A.B. Spierings, D. Mohr, Additively manufactured metallic microlattice materials for high specific energy absorption under static and dynamic loading. Acta Mater. **116**, 14–28 (2016)

S. Tsopanos, R.A.W. Mines, S. McKown, et al., The influence of processing parameters on the mechanical properties of selective laser melted stainless steel microlattice structures. J. Manuf. Sci. Eng. (ASME) **132**, 041011-1–12 (2010)

I. Ullah, J. Elambasseril, M. Brandt et al., Performance of bio inspired Kagome truss core structures under compression and shear loading. Comp. Struct. **118**, 294–302 (2014)

R. Vrana, D. Koutny, D. Palousek, Impact resistance of different types of lattice structures manufactured by SLM. Mod. Mach. Sci. J. 1579–1585 (2016)

R. Vrana, D. Koutny, D. Palousek, et al., Impact resistance of lattice structure made by selective laser melting from AlSi12 alloy. Mod. Mach. Sci. J. 852–855 (2015)

R.E. Winter, M. Cotton, E.J. Harris et al., Plate impact loading of cellular structures formed by selective laser melting. Model. Simul. Mat. Sci. Eng. **22**(025021), 1–23 (2014)

L. Yang, O.A. Harrysson, H.A. West II et al., Low energy drop weight performance of cellular sandwich panels. Rapid. Prot. J. **21**(4), 433–442 (2015)

Chapter 8
Conclusions from the Book: Themes, Future Research Strategies

Abstract Five themes were identified in the preface, and the conclusions on these are now given. The five themes where enhancement in specific structural performance, use of existing structural theories, integration of realisation methods, the general field of architectured materials and formal computer based methods. Also, eight possible areas for further research are identified. Finally the alternative perspective of exploring the *design (property) space* of architectured cellular materials is discussed, from the specific design point of SLM BCC SS316L, discussed in detail here. The book represents a focussed subset of the wide ranging and fast developing fields of architectured (architected) cellular materials, mechanical metamaterials, hierarchical metamaterials, and multi functional materials.

Keywords Architectured materials · Architected materials · Standardisation · Certification · Design space

8.1 The Five Themes

8.1.1 The Enhancement of Specific Structural Performance Using Additive Manufacturing

It has been shown that significant improvements in stiffness and strength of beams, in foreign object impact performance of panels and in the performance of energy absorbing devices can be achieved without full optimisation of the microlattice and with conventional structural solutions. Improvements have been quantified and these can be achieved by enhancing topology (dimension, geometry), processing parameters and parent material. Improvements in structural performance of up to 500% can be achieved. These improvements are dependent on enhancing the quality of the final additively manufactured structure. There is scope for further optimisation of the microlattices using formal methods and the development of non conventional structural solutions.

R. Mines, *Metallic Microlattice Structures*, SpringerBriefs
in Structural Mechanics, https://doi.org/10.1007/978-3-030-15232-1_8

8.1.2   Integration of Manufacturing, Materials and Application

Additive manufacturing is a fast developing enabling technology. Issues to address include full exploitation of additive manufacturing, full definition of quality, definition of advantages and disadvantages of various additive manufacturing processes, and full control (and optimisation) of the realisation process. These will depend on structural scale. Hence, to achieve the required structural solution, a multi disciplinary approach is required. New additive manufacturing processes will appear in the future, which may change structural and process solutions. A current process under development is binder jetting, which has the potential for wide industrial application and a wide range in scale. New materials, specifically designed for additive manufacturing, are coming on stream for enhanced properties. The additive manufacturing processing profile can be adjusted to specifically enhance existing material properties.

8.1.3   Existing and Developing Structural Theories

The structural applications discussed in this book cover a wide range of non linear material and structural behaviour. This includes large strain plasticity, material rupture, geometric non linearity, and impact (low to high rate) loading. These topics have been addressed for many years for conventional structures. Given the complexity of structures produced by additive manufacturing, these conventional theories need to be adapted and enhanced to fully exploit the potential of Additive Manufacturing Technology. Specific examples of structural impact further research as identified by Sun and Li (2018) include:

- Base material behaviour (and influence of manufacturing process)
- Detailed modelling of different impact regimes
- Study of rupture behaviour
- Study of cellular stress wave behaviour (dependent on topology and elastic properties).

A note of caution from experience: in depth study and characterisation of Additively-Manufactured microlattice structures are only effective when the realisation parameters are defined, controlled and stable.

It should be noted that additive manufacture can be extended to macro scale. For example, Buchanan et al. (2017) constructed stainless steel 316L hollow (stub) columns with typical cross sections of $50 \times 50 \times 5$ mm thick using direct metal laser sintering. They compared additively manufactured produced columns with conventionally produced (wrought) columns, and showed compression behaviours were similar. This is believed to be one of the first applications of metal additive manufactured to macro structures. It would be interesting to realise the conventional microlattice structures discussed in Chap. 2, with equivalent structures manufacture using additive manufacture. Build envelopes in powder bed fusion machines are increasing.

8.1.4 Specific Issues Versus the General Field (Architectured Cellular Materials, and Industrial Processes)

A set of specific state of affairs has been considered in this book, namely restricted manufacturing (SLM, EBM, PPWG, WM, Binder Jet), restricted (metal) materials (stainless steel 316L, titanium alloy Ti 64, aluminium alloy AlSi10/12Mg), restricted topology and scale (microlattice, BCC, BCCZ, OT) and restricted structural applications (beam, panel, energy absorbers). These specific aspects are part of the larger fields of Architectured (Architected) Cellular Materials. Schaedler and Carter (2016) identified 13 additive manufacturing processes from micro to nano scale, and for various materials. They discussed lightweight structures, energy absorbing structures, metamaterials, materials for thermal management, and materials for bio scaffolds. The latter have not been discussed here, but it should be noted that the original SLM research at the University of Liverpool was motivated by bio implants (see Chap. 3).

Extension of metallic lattice structures to nano scale was discussed in Chap. 3, with binder jetting and materials jetting realisation techniques. Nano lattices have not been discussed in detail in this book as these tend to be made from polymer and ceramic materials, and they address different structural applications. Papers by Valdevit et al. (2011) and Meza et al. (2017) provided good starting points for the research fields. Valdevit et al. (2011) addressed stiffness and strength behaviours, as well as multi function and optimal design. Meza et al. (2017) systematically investigated modulus and strength for nano lattices, with associated scaling laws. They concluded that the classification of nano lattices as bending and stretch dominated is insufficient, and new theories are required to include nodal effects.

Extension of the specific issues include:

- Formally include multi functionality
- Extend from lattices to minimal surfaces
- Quantified effects of scaling micro to nano including large strain plasticity and rupture
- Develop optimisation methods for nonlinear behaviour
- Ensure full control and optimisation of the realisation process.

Another starting point is Valdevit et al. (2018), and the associated Journal of Materials Research special issue on architected materials.

The control, optimisation and standardisation of the complete additive manufacturing process is being considered across industries. As selective laser melting and electron beam melting become more industrially mainstream, so the realisation process and quality of the final component needs to be fully defined. The control and optimisation of the selective laser melting and electron beam melting processes were discussed in Chap. 3.

Structural applications discussed here have been specific e.g. beams, panels, energy absorbing devices, however this focus is part of the larger activity of realising industrially relevant components and products using 3d printing. Hence, formal standards and methodologies need to be developed. An example industrial methodology is from Lloyds Register and TWI (2017). This is based on UK certification activity and is based on BS EN 17296 standards on additive manufacturing (British Standards 2016).The guidance notes cover: definition of additive manufacturing processes, suitability of additive manufacturing processes for a given application, design of additive manufacturing produced parts, provision and use of additive manufactured materials, additive manufacturing methods, post processing inspection and testing, and health and safety management methods (documentation, organisation, intellectual property). The ultimate aim is to certify the complete realisation process.

8.1.5   Fundamental Ideas Underlying Computer Based Realisation Methods

Full exploitation of additive manufacturing for microlattice structures is increasingly based on sophisticated computer based methods. These address full control and optimisation of the realisation process, and synthesis of tailored and optimal structural solutions. The structural applications addressed in this book are nonlinear and dynamic, hence synthesis and optimisation methods are still under development. Matters are further complicated when multi functionality is addressed. The specific case of manufacturing selective laser melted microlattice structures was discussed in Chap. 3 (Sing et al. 2018 and Mcmillan et al. 2017).

Smith et al. (2016) split additive manufacturing realisation activities into 1. Process to performance, 2. Materials, 3. Experimental methods and 4. Computation, mathematics and data informatics. As far as process to performance is concerned, the aim is to optimise process and material micro structure. As far as materials are concerned, the aim is to optimise materials usage, material properties using modelling and materials databases. These activities should be supported by experimental characterisation and process monitoring. Finally, this activity should be informed and supported by databases, modelling and simulation, data mining and applied mathematical methods. Hence, the activities and issues discussed in this book should be integrated into the formal structure given previously. This formalisation is being addressed by standardisation and certification methodologies.

Yan et al. (2018) developed a process structure property modelling framework for additively-manufactured components realised using electron beam melting, titanium alloy Ti 64, and for component fatigue behaviour. The additive manufacturing modelling framework consisted of powder dynamics, thermal and computational fluid dynamics models, grain growth and fatigue prediction. The authors validated their approach and investigated the effect of various realisation parameters on fatigue performance. The authors identified additions to the framework, namely: CAD tools, machine control, post processing and topology optimisation.

Garanger et al. (2017) discussed a formal methodology for full specification and control of the additive manufacturing processes. They commented that further work is required to physically understand the additive manufacturing process.

From a structural design point of view, Chap. 5 discussed sophisticated synthesis and optimisation methods. For example, Chen et al. (2018) used Isomap and Voxel techniques to develop extremal cellular structures. Also, for multi functional behaviour, the different behaviours need to be formalised into a single model, in order to develop a coherent optimum (e.g. Ro and Roper 2017). Hence, it can be concluded that computer based methods will become increasingly complex for all aspects of the realisation process, and ultimately a fully defined realisation methodology will be developed.

8.2 Some Suggestions for Future Research

An aim of this book has been to give an overview of the entire field. In this way, the context of specific research has been highlighted. A number of research activities can now be identified:

1. Now that metallic lattices can be made at nano scale (Saleh et al. 2017 and Vyatskikh et al. 2018), there is scope to transfer ideas between micro and nano scale (See Chap. 3). Two issues for nano scale are expanding possible metals and effect of scale on plasticity and rupture. Impact testing at near nano scale is being developed. For example, Lai and Dario (2018) developed a small pneumatic gun to test microlattices with strut diameter of five microns and block size of $9 \times 9 \times 12 \ \mu m$. The impact velocity was between 0.15 and 0.25 ms^{-1}. Also there is a need to develop multi scale materials, with micro and nano features, to further enhance functional performance.

2. There is scope to expand metallic additive manufacture into macro (conventional structure) scale (Buchanan et al. 2017). It would be interesting to revisit conventional lattice structures (Chap. 2), and apply the additive manufacturing processes.

3. We need to distinguish between design methodologies and research methodologies. Design methods in the main address stiffness and strength of complex lattice structures (Chap. 5), but cannot as yet deal with transient nonlinear problems (Osanov and Guest 2016). Hence, research methods need to address heuristic methods and effective exploration of the *design space*. A specific research topic relevant to the applications discussed here is nonlinear optimisation.

4. Additively-Manufactured lattices are being extended to minimal surfaces and shell structures (Al Ketan and Al Rub 2018). Shell structures have also been

addressed by Bonatti and Mohr 2017. Shell structures dissipate stress concentrations and give further collapse modes (see Chap. 7).

5. The structural researcher needs to keep up with developments in additive manufacturing processes. For example, selective laser melting is now being specifically focused on micro scale and the build envelope is increasing. Also, the processes are being monitored and controlled, and SLM Machines are coming down in price. Binder Jetting (Material Jetting, Bind and Sinter) approaches need to be fully validated for microlattice structures. Hence, the structures researcher needs to fully define, control and optimise the additive manufacturing realisation process prior to structures research.

6. Three aspects of Materials Technology are being addressed, namely: process optimisation of existing materials (Liu et al. 2017), development of new materials and development of detailed constitutive relations. e.g. effect of scale on plasticity and rupture (Wang and Li 2018) (See Chaps. 3 and 4).

7. This book has addressed a small number of specific structural problems, and this needs to be extended to multi functionality (Xiong et al. 2015). For example, structural performance, fluid transfer and fluid flow can be modelled in single analytic functions (Ro and Roper 2017), so that multi functional behaviour can be formally optimised. Multi functionality for additive manufacturing metals is currently less well developed as compared to additively-manufactured polymers (Yang et al. 2018). Yang et al. (2018) gave a review of multi functional structures and they cover single materials, multiple materials, shear assisted additive manufacture, magnetic assisted additive manufacture, electric and acoustic associated additive manufacture, shape changing in structures, hydrodynamic surface structures, microfluidic structures, optical devices, transparent glass, biological structures, multi scale structures and electronic devices. This book has focused on single functions, on single materials and on a single scale. As far as the materials and processes discussed in this book are concerned, the processes of binding jetting and material jetting are most suitable for inclusion of multi materials, which can be seen as a specific next step (see Chap. 3).

8. Finally there are large number of structural impact problems to be addressed (Sun and Li 2018). These cover low and high speed impact, as well as blast loading. Gümrük et al. (2018a) have addressed material strain rate and micro inertial issues for a simple microlattice loading.

8.3   An Alternative Approach: The Investigation of the *Design (Property) Space* for Selected Structural Applications

An alternative approach to the field is using the concept of 'design space'. The first design point considered in this book was SLM—SS316 L—BCC—2.5 mm, where the latter is the cell size. This design point was justified from the points of view of ease of manufacture and appropriateness to the structural applications under consideration. Exploration away from the 'design point' was then considered for different cell sizes (1–2.5 mm), different materials (Ti 64, Al Alloy), different topologies (BCC, BCCZ), for electroless plating and for impact loading.

Table 8.1 summarises specific collapse stress (σ_c/ρ) and density properties, and Fig. 8.1 plots $\log(\sigma_c/\rho)$ versus relative density.

As far as the University of Liverpool results are concerned, the effect of reducing cell size from 2.5 to 1.5 mm is to increase crush load and density. The effects of impact loading, electroless plating and adding Z struts are to increase specific strength. Finally the effect of using Ti 64 is to increase specific strength and density. It was noted in Chap. 4 that the Ti 64 parent material properties were not optimal, and hence compression collapse properties could be improved. All collapse modes are similar.

The Kagome topology is optimized in one direction and has a large strut diameter, and strength and density properties are good. The maraging steel BCC and shell have high relative density, but have high specific strength. In the case of BCC, maraging steel yield stress is greater than that for SS316L, and the overall quality of the lattice is improved over SLM BCC. The improvement of performance for the shell structure over BCC for the maraging steel can be seen.

Finally the nanolattice has high relative density, but specific strength is comparable with the University of Liverpool work.

Hence, Fig. 8.1 illustrates improvements away from the control point of SS316L—SLM—2.5 mm—BCC. The figure illustrates the various strategies for improving structural performance. The data is by no means comprehensive, which shows the effort required to fully populate the *design (property) space*.

8.4   Overall Conclusions from the Book

The overall aim of this book has been to capture the fast developing technology enabling the use of additive manufacturing to realise improved metallic cellular (lattice) structures. The focus has been on specific structural performances (e.g. beam, panel, energy absorbing devices), and on specific periodic cellular solutions. The aim has been to define and clarify these aspects, before tackling more open ended issues relating to generalised cellular design and to mimicking biological systems.

The structural performance considered here has been non linear and dynamic, hence complete formal synthesis methods are, as yet, not possible and heuristic approaches are required. In order to effectively exploit Additive Manufacturing Technology, an integrated approach (manufacturing—materials—application) is required. This is being formalised in next generation additive manufacturing realisation systems.

This book has mainly addressed short to medium technology, viz powder bed fusion processes and conventional materials. There is certainly more to exploit with these technologies, especially by formalising and simulating the complete realisation process. However, in the intermediate and long-term, new realisation processes will

Table 8.1 Comparison of compression specific collapse load for lattice with relative density. S—static. Density SS316L/maraging steel = 7860 kgm^{-3}, density Ti 64 = 4680 kgm^{-3}, density Ni = 8910 kgm^{-3}

Source	Gümrük and Mines (2013)	Gümrük and Mines (2013)	Gümrük et al. (2013)	Gümrük et al. (2018b)	Gümrük et al. (2018a)	Mines et al. (2013)	Ullah et al. (2016)	Vyatskikh et al. (2018)	Al Ketan and Al Rub (2018)	Al Ketan and Al Rub (2018)
	A	B	C	D	E	F	G	H	I	J
Parent material	SS316L	SS316L	SS316L	SS316L/Ni	SS316L	Ti 64	Ti 64	Ni	Maraging steel	Maraging steel
Topology	BCC	BCC	BCCZ	BCC	BCC	BCC	Kagome	BCC	BCC	TPMS shell
Cell size (mm)	2.5	1.25	2.5	2.5	2.5	2.5	11.5	2.5×10^{-3}	10	10
Loading	S	S	S	S	Impact	S	S	S	S	S
Electroless plated	–	–	–	Yes	–	–	–	–	–	–
Strut diameter (μm)	210	210	210	244	210	372	810	0.42	~1000	~500
Lattice density (kgm^{-3})	288	990	296	417	288	285	50	3300	1415	1415
Relative density	0.037	0.135	0.038	0.053	0.037	0.061	0.011	0.37	0.18	0.18
Block collapse stress (MPa)	0.430	5.373	1.464	0.800	0.645	2.000	3.900	17	17.0	50.0
Specific block collapse stress (Pa kg^{-1}m^3)	1493	5427	4946	1918	2240	7017	78,000	3030	12,014	35,336
Log (specific block collapse stress)	3.174	3.735	3.694	3.283	3.350	3.846	4.892	3.481	4.080	4.548

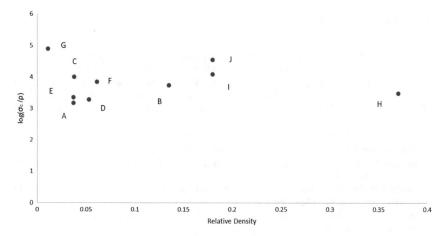

Fig. 8.1 Lattice block compression: log (log(σ_c/ρ)) versus relative density for University of Liverpool results (**a–f**), Kagome (**g**), nano lattice (**h**) and surface lattice (**i, j**). For definition of **a–j** see Table 8.1

appear, which may build on existing processes or maybe new processes. Schaedler and Carter (2016) identified 13 additive manufacture processes and ranked them in the level of maturity from 1 for most mature to 13 for least mature. They identified selective laser melting as 2, electron beam melting as 3, binder jetting as 6, and photo polymer wave guides as 10. Hence it can be concluded that selective laser melting and electron beam melting are mature industrial processes that are subject to industrial certification, but are stable enough for in depth engineering science study.

It should be noted that this book has, in the main, addressed single materials, single functions and a single (micro) scale. Activities could usefully be extended to nano scale, to multi functionality and to more innovative (multiple) metallic materials. However, by focusing on conventional additive manufacturing processes and materials, the aim has been to give systematic foundations to underlying structural and materials aspects for selected applications using current state of the art additive manufacture. However, where possible the content of the book has been related to the more wide ranging and fast developing fields of architectured (architected) cellular materials (Schaedler and Carter 2016), mechanical metamaterials (Yu et al. 2018), hierarchical metamaterials (Zheng et al. 2016) and multi-function materials (Yang et al. 2018).

Hopefully, the book has identified some of the major issues that need to be considered by the researcher in the field and has identified the issues that will be further developed by the additive manufacturing and structures research community in the near future.

References

O. Al Ketan, R.K.A. Al Rub, The effect of architecture on the mechanical properties of cellular structures based on the IWP minimal surface. J. Mater. Res. **33**(2), 343–359 (2018)

C. Bonatti, D. Mohr, Large deformation response of additively manufactured FCC metamaterials: from octet truss lattices towards continuous shell meso structures. Int. J. Plast. **92**, 122–147 (2017)

British Standards, Additive manufacture: general principles. BS EN ISO 17296: 2016 Parts 1–4 C4 (2016)

C. Buchanan, V.P. Matilainen, A. Salminen et al., Structural performance of additive manufactured metallic material and cross sections. J. Constr. Steel Res. **136**, 35–48 (2017)

D. Chen, M. Skouras, B. Zhu, et al., Computational discovery of extremal microstructure families. Sci. Adv. **4**, eaao7005 (2018)

K. Garanger, E. Feron, P.L. Garoche, et al., Foundations of intelligent additive manufacturing, 12 May 2017 (2017), arXiv:1705.00960v2

R. Gümrük, R.A.W. Mines, Compressive behaviour of stainless steel microlattice structures. Int. J. Mech. Sc. **68**, 125–139 (2013)

R. Gümrük, R.A.W. Mines, S. Karadeniz, Static mechanical behaviour of stainless steel microlattice structures under different loading conditions. Mat. Sci. Eng. **A586**, 392–406 (2013)

R. Gümrük, R.A.W. Mines, S. Karadeniz, Determination of strain-rate sensitivity of microstruts manufactured using selective laser melting method. J. Mater. Eng. Perf. (ASM) **27**(3), 1016–1032 (2018a)

R. Gümrük, A. Usun, R.A.W. Mines, The enhancement of the mechanical performance of stainless steel microlattice structures using electroless plated nickel coatings. MDPI Proc. **2**(8), 494 (2018b)

C.Q. Lai, C. Daraio, Highly porous microlattices as ultra thin and efficient impact absorbers. Int. J. Imp. Eng. **120**, 138–149 (2018)

L. Liu, Q. Ding, Y. Zhong et al., Dislocation network in additive manufactured steel breaks strength-—ductility trade off. Mater. Today **21**(4), 354–361 (2017)

Lloyds Register/TWI, Guidance notes for certification of metal parts made by additive manufacturing (2017)

M. McMillan, M. Leary, M. Brandt, Computationally efficient finite difference method for metallic additive manufacturing: a reduced order DFAM tool applied to SLM. Mater. Des. **132**, 226–243 (2017)

L.R. Meza, G.P. Philipot, C.M. Portela et al., Re examining the mechanical property space of three dimensional lattice architectures. Acta Mater. **140**, 424–432 (2017)

M. Osanov, J.K. Guest, Topology optimisation for architected materials design. Ann. Rev. Mater. Res. **46**, 211–233 (2016)

C.J. Ro, C.S. Roper, Analytical models of the geometric properties of solid and hollow architected lattice cellular materials. J. Mat. Res. **33**(3), 264–273 (2017)

M.S. Saleh, C. Hu, R. Panat, Three dimensional micro architected materials and devices using nano particle assembly by pointwise spatial printing. Sci. Adv. **3**, e1601986 (2017)

T.A. Schaedler, W.B. Carter, Architected cellular materials. Ann. Rev. Mater. Res. **46**, 187–210 (2016)

S.L. Sing, F.E. Wiria, W.Y. Yeong, Selective laser melting of lattice structures: a statistical approach to manufacturability and mechanical behaviour. Robot. Comput. Integr. Manuf. **49**, 170–180 (2018)

J. Smith, W. Xiong, W. Yan et al., Linking process, structure, property and performance for metal based additive manufacturing: computational approaches with experimental support. Comp. Mech. **57**(4), 583–610 (2016)

Y. Sun, Q.M. Li, Dynamic compressive behaviour of cellular materials: a review of phenomena, mechanism and modelling. Int. J. Imp. Eng. **112**, 74–115 (2018)

I. Ullah, M. Brandt, S. Feih, Failure and energy absorption characteristics of advanced 3D truss core structures. Mater. Des. **92**, 937–948 (2016)

L. Valdevit, K. Bertoldi, J. Guest, et al., Architected materials: synthesis, characterisation, modelling and optimal design. Introduc. Spec. Issue J. Mater. Res. **33**(3), 241–246 (2018)

L. Valdevit, A.J. Jacobsen, J.R. Greer, et al., Protocols for the optimal design of multi functional cellular structures: from hypersonic to micro architected materials. J. Am. Ceram. Soc. **94**(S1), 1–20 (2011)

A. Vyatskikh, S. Delalande, A. Kudo et al., Additive manufacturing of 3D nano architected metals. Nat. Commun. **9**(593), 1–8 (2018)

Z. Wang, P. Li, Characterisation and constitutive model of tensile properties of selective laser melted Ti6Al4V struts for microlattice structures. Mater. Sci. Eng. A. **725**, 350–358 (2018)

J. Xiong, R.A.W. Mines, R. Ghosh et al., Advanced microlattice materials. Adv. Eng. Mater. **17**(9), 1253–1264 (2015)

W. Yan, Y. Lian, C. Yu et al., An integrated process-structure-property modelling framework for additive manufacturing. Comput. Meth. Appl. Mech. Eng. **339**, 184–204 (2018)

Y. Yang, X. Song, X. Li et al., Recent progress in bio mimetic additive manufacturing technology: from materials to function structures. Adv. Mater. **30**(36), 1706539 (2018)

X. Yu, J. Zhou, H. Liang et al., Mechanical metamaterials associated with stiffness, rigidity and compressibility: a brief review. Prog. Mater. Sci. **94**, 114–173 (2018)

X. Zheng, W. Smith, J. Jackson et al., Multiscale metallic metamaterials. Nat. Mater. **15**, 1100–1106 (2016)

Printed in the United States
By Bookmasters